WHEN MACHINES BECOME CUSTOMERS

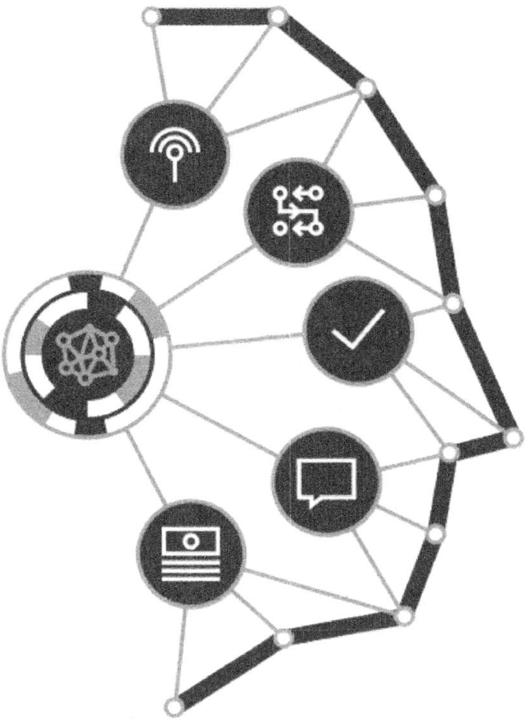

WHEN MACHINES BECOME CUSTOMERS

Don Scheibenreif

Mark Raskino

Cover Art: Walter C. Baumann

Editor: Bennett Voyles

3rd Edition – August 2025

First published January 2023 by
Gartner Inc. 56 Top Gallant Road, Stamford, CT USA

Contact: books@gartner.com

ISBN 979-8-9890801-4-4

For my wife Mona

Don Scheibenreif

For my mother Audrey

Mark Raskino

And for the late Gideon Gartner
from us both.

Contents

Foreword

Preface

1. Billions of Brand-New Customers Are Headed Your Way ... 6
2. Meet Your New Machine Customers ... 28
3. The Custobots ARE Coming ... 48
4. Machine Customers Could Make or Break Your Business ... 66
5. Machine Customers Are Made, Not Born – Who'll Be the Makers? ... 92
6. Determining The Relevance of Machine Customers To Your Situation ... 112
7. How Machine Customers Work (in a Generative AI world) ... 134
8. How Machine Customers Could Hurt Your Business ... 172
9. Frenemy 2.0: What Machine Customers Will Mean for You Personally ... 186
10. The Lasting Impact of Machine Customers - Economy and Business ... 202
11. Where Machine Customers Will Lead Us - The Long-Term Future ... 220

Notes ... 238

Index ... 254

Acknowledgments ... 262

About the Authors ... 264

Other Gartner Books ... 265

3rd Edition - Updated for Agentic AI

Technology continues to move at dizzying speeds. The first edition of this book was finalized in November 2022 – just before the seismic shockwave called ChatGPT. The second edition was published in September 2023 to capture the emerging impact of Generative AI. The creative explosion that accompanied its arrival is set to be significantly amplified by the developments in Agentic AI. Both are accelerating the advancement of machine customers even faster than we had imagined. Additionally, a book like this always sparks many questions and conversations with readers, from which more ideas and insights flow. In addition to new Gartner research on machine customers, we have had a lot of dialogue with Gartner clients, conference attendees and in forums like LinkedIn. As with our second edition, this new August 2025 third edition includes some of these important insights and additions to our thinking. Specifically:

- An expanded discussion of the Machine Customers Decision Path to help readers pick a strategy that is right for their organization (c. 6)
- Expanded discussion of how business functions like Sales, Marketing, and Customer Service will need to adapt in a world of Machine Customers (c. 6)
- The implications of Agentic AI on Machine Customer development (c.7)
- Expanded discussion on the AI techniques and other technologies needed to power Machine Customers (c. 7)
- Expanded discussion on Data Management, Identity, Security and Risk Management in an AI world (c.7)
- The introduction of Gartner's emerging research on Autonomous Business and its connection to Machine Customers (c.11)
- New Generative AI case examples of emerging Machine Customers (throughout)

Foreword

Many creative collaborations start with a simple question: "What if?" More than a decade ago, I was in a Gartner office in Stamford, Connecticut with analysts Jenny Sussin and Don Scheibenreif. Analysts are often solitary, working on research in their home offices and taking video calls with clients around the world. But this day, we were together; Jenny with her welcoming hug and Don with his million-dollar smile. As we were chatting about customer experience I said, "what if the customer is a machine?". That seed was planted firmly, and intentionally, into fertile ground.

Jenny went on to manage analysts covering customer service while Don took the idea of Machine Customers and let it blossom. His deep background in consumer marketing and digital business was the perfect proving ground. Eventually he began collaborating on the concept with Mark Raskino, a brilliant Gartner analyst always up for creating a grounded view of the future.

The first edition of this book was born a few years later and collected the best visions of a world where machines were becoming customers. Then the Generative AI boom began in November 2022. Well, that required a quick second edition to include perspective on GenAI, this eerily human-like capability that will play a larger and larger role in multiple user experiences. AI is quickly evolving to include agents which can autonomously tackle problems, assemble and optimize solutions, remember, learn, and reason. Agents will function as customers on their own, and as brokers for humans to help navigate the complexities of a fully digital world. And so this work from Don and Mark also evolves.

I am so fortunate to have been in the creative flow that day in Jenny's office. I know this book will lead you to your own "What Ifs?".

Chris Howard
Chief of Research, Gartner
June 2025, New Hartford Connecticut.

Preface

A business growth megatrend is quietly taking shape that we believe will be more significant than the arrival of digital commerce. For the first time in history, companies will be able to make their own customers - agents powered by artificial intelligence and other technologies that can shop on behalf of individual people, organizations and even for themselves. These new buyers – we call them "custobots" – will soon create mega markets. CEOs we have polled believe that, on average, 15 to 20% of their company's revenue will come from Machine Customers by 2030. Gartner modeling predicts Machine Customers will be directly involved in or have influence over many trillions of dollars in purchases, given the rapid advancements of Generative AI.
With *three* varieties of customers in play - consumers, businesses, and now machines – all types of enterprises will find new revenue opportunities. In many cases, the pie will get bigger. New markets will open that were otherwise closed. But capturing this growth will require new ways of managing customer relationships and new business models. Machine Customers will change how you operate and who you hire. Success will demand that you let go of many old practices and habits of thought. The underlying research for this book has evolved over most of a decade. When we compared what we do as human customers with the emerging capabilities of the Internet of Things (IoT) and intelligent systems (AI), it became obvious that machines can perform some customer 'work' as well as, or better than humans. We wrote this book for business leaders who must consider big trends that disrupt existing strategies or create significant growth opportunities - CEOs, strategy officers, technology officers (CIO, CDO, CTO), marketing officers, sales officers, and supply chain officers. Other professionals, academics, and students may also find value in it. Come with us as we challenge what it means to be in business when your customers are machines. How Machine Customers will make our lives better – mostly. How they will change business and operating models, open some doors and close others - as they transform marketing, sales, and virtually every company. Psychologists sometimes suggest that we don't have ideas; ideas have us. The Machine Customers concept is like that. Once it seized us, there was no going back. There's still time for you to close this book before your preconceptions about customers and markets are shattered. Turn the page and the idea virus will be in you too.

Don Scheibenreif and Mark Raskino

Chapter 1: Overview

Billions of Brand-New Customers Are Headed Your Way

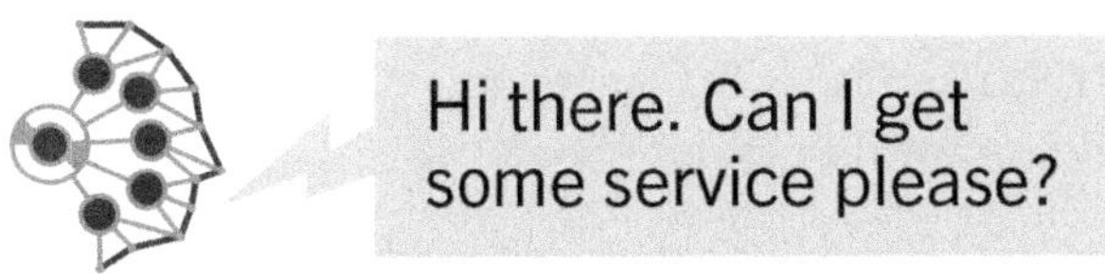

Soon, intelligent machines will not only be workers - they will also be customers.

"We love our customers!" gushes your corporate website. Really? Let's be honest; your customers are a pain. They don't understand your product, they don't pay attention, they don't turn up on time, they haggle irrationally, and then they forget to complete the sales process. Much of the time, it's like dealing with a wayward group of easily distracted children with short attention spans. It's not their fault - what you sell isn't particularly exciting for many of them. And from their perspective, being a customer takes work. But there's no other choice, right? You just have to deal with them.

That has been true for the entire history of business.

But soon, for the first time, it might not be.

Because your customer no longer needs to be a person.

Welcome To a World Where Your Best Customers May Not Be Human

Let's peer into that future for a moment and catch up with Jessica. It is sometime in the early 2030s, and Jessica is an ambitious, hardworking local journalist on her way to work.

"Jessica, the batteries are at 10%. I booked a recharge – we'll stop in three miles. Eight minutes, twenty dollars, arrival time unchanged". She didn't acknowledge it - no need to. But the car's voice had interrupted her train of thought, so she might as well check messages. The first was from her oven, noting it had booked a gig worker to clean it. Last weekend's roast duck was messy, but SO worth it. "Alexa - reorder the duck" she said. "I will upgrade to organic, it's almost the same price, and other customers prefer it." replied the cloud-based virtual assistant. Jessica liked the sound of going organic, so she did not overrule the custobot. Her editor was next, pestering her for a story, so she checked on the PIXchange news media NFT marketplace. A train station AI security camera offered a video titled 'anomalous-crowd-behavior-level3'.

People were lying on the floor… it looked like a protest. Good story! In seconds, her virtual picture editor haggled for a blockchain property-rights assured clip from the camera for $300. "Hey, car reroute to the station after we charge up, please." "OK, Jessica." "There's a wheelbot at the station. I can replace a tire while we charge." Just before the car turned into the service station, Jessica paused for a moment. She remembered what things were like before all these custobots became part of her routine, and the fear people had initially felt about delegating to them. Now nobody could live without them.

A scenario like this is less a possibility than an inevitability. It's a glimpse of a future in which billions of machines will become customers, initiating a wave of economic growth in the 2020s and 30s that could compare to the arrival of the emerging markets in the 1990s. In fact, we predict that by 2028, a Machine Customer will earn Customer of the Year at a global top 500 Company.

We know this sounds a bit futuristic. Your first thoughts might be about Sentient AI, such as Skynet, HAL and a bunch of other killer robots. That is normal. The idea of turning over an essentially human activity to a machine gives all of us pause. But didn't we all have similar misgivings once upon a time about e-commerce, smart speakers, and the cloud? Each time, skepticism gave way to curiosity, experimentation, and acceptance, as an idea that once seemed science fiction gradually became part of our lives. We believe Machine Customers have the same potential.

Come with us on a journey as we challenge what it means to be a customer. Soon, being a customer will no longer be the exclusive purview of a human being. Being a customer will take on several forms, and Machine Customers will take a lot of drudgery out of our lives and open new horizons for growth.

Come with us as we challenge what it means to be in business when your customers are machines. How Machine Customers will make our lives better – at least for the most part. How they will change business and operating models, open some doors and close others - as they transform marketing, sales, and virtually every business function. We promise to show you that this future is real, it is happening today, and there is something good in it for you.

Let's get started.

What Is a Machine Customer Exactly?

A Machine Customer is a kind of digital agent that acts on our behalf. A Machine Customer could be an AI-enabled virtual assistant like Siri or Alexa or a physical object connected to the internet, like your car, your washing machine, or a factory robot. We will stay in control, but now we have our own concierge who will do a lot of the tedious work of being customer for us.

Definition: A Machine Customer is a non-human economic actor who obtains goods or services in exchange for payment

---

The authors first realized that Machine Customers were a viable concept when we compared what we do as human customers with the emerging capabilities of the Internet of Things (IoT) and intelligent systems (AI). It became obvious to us that machines can perform the same tasks we do as human customers and, in some cases, do them better:

- Receive messages from businesses trying to sell to us.
- Search for information.
- Negotiate for the best deal.

- Buy goods or services.
- Request service when something goes wrong.
- Tell others about our experiences.

Today, no single machine can perform all of these tasks, but we have dozens of emerging examples of machines that perform at least some. The evolution of machines becoming customers is well underway. As Aaron Rajan, Global VP of Consumer Digital Experience at Unilever told us: "I see a future where algorithms will increasingly serve to represent our interests online, to anticipate what you will need to discover and seek out new experiences on your behalf."[1] Beyond that, the authors believe that Machine Customers will eventually focus not just on you, but also on their own needs.

We believe the era of Machine Customers will evolve in three main phases over the next 15 years:

Bound Machine Customers (today)

At first, humans will set precise rules for the machine to follow. The machine is "bound" to a single supplier of products or services and is responsible for ensuring the decision is executed. This is happening today as traditional products are remastered with AI-driven digital services. Early examples include HP Instant Ink, Amazon Alexa, Tesla vehicles, and Autocrib's Robocrib (an internet-enabled industrial vending machine).

Adaptable Machine Customers (mid-term)

In this phase, the machine breaks out of a "walled garden" and can choose among competing options. In this case, the human may state an end goal to a digital assistant, such as: "Alexa, we need more laundry detergent" and the machine will have some discretion and autonomy on what brands to buy, at what price, and from which supplier. The delegation of process steps is still broadly human-defined, and human intervention is still required for some parts of the process, but the machine can adapt. Early examples include robo-trading, ethernet auto-negotiation, Staples Easy Ordering System, and robo-advisors like Betterment and Wealthfront.

Autonomous Machine Customers (longer term)

In this third and most advanced phase, the Machine Customer can act independently on behalf of human customers. It can operate with a high degree of discretion and owns most of the process steps leading up to and following a transaction. Eventually, machines will also act on their own behalf to serve some of their own needs. Early examples of this level of advancement are few, but we see their potential by extrapolating from some of the most recent advances in artificial intelligence, especially generative AI and agentic AI technologies explain in Chapter 7.

Why Should You Care About Machine Customers?

Growth. Growth. Growth. It's simple - the Machine Customer era will be a business growth megatrend more significant than the arrival of digital commerce. CEOs are looking for growth wherever they can find it. Growth comes from innovation. It creates momentum and attracts investment. Growth feeds on itself. However, sometimes no matter what they do, companies find it increasingly difficult to grow. That's where Machine Customers come in.

Here's why: in the past, companies typically had to spend a lot of time finding and keeping customers. The age of Machine Customers will change that, by creating megamarkets, mega fast. For the first time in human history, companies will be able to manufacture their own customers. CEOs we have polled believe that, on average, 15 to 20% of their company's revenue will come from Machine Customers by 2030. And 50% of CEOs will have a strategy for machine customers by 2026-7[2]. Gartner modeling predicts Machine Customers will be directly involved or have influence over a cumulative $30 trillion in purchases over a decade. Gartner modeling predicts Machine Customers will be directly involved or have influence over a cumulative $30 trillion in purchases over a decade.

With three varieties of customers - consumers, businesses, and machines– all types of enterprises will find new revenue opportunities. In many cases, the pie will get bigger. New markets will open that were otherwise closed. But capturing this growth will require new ways of managing customer relationships and new business models. Machine Customers will change how you operate and even who you hire. Success will demand that you let go of some old practices and habits of thought, to make way for this new world.

Some simple, centralized corporate automatic replenishment systems have existed for many years. Now, that capability is about to be democratized and diffused. By 2027 50% of people in advanced economies will have AI personal assistants working for them daily[3], and it has already started.

By 2030 there will be at least 7 billion enterprise B2B IoT enabled devices with the potential to be smart enough to behave as customers[4] that can shop for services and supplies for their owners and themselves. The number of B2C devices is expected to reach 4 billion by 2030[5]. Virtual AI assistants or AI agents are also expected to number in the billions. As these billions of Internet-connected physical products and ChatGPT style AIs improve, they will be able to discover, compare, negotiate, buy and sell. Internet-connected, with delegated shopping and purchasing rights, they will become a new type of customer that acts on our behalf. We will let them shop for us at home and at work because they will make more intelligent, more timely purchasing choices. That will happen because, let's face it, buying toilet tissue, pet food, and life insurance are chores that eat your life - not fun leisure experiences.

In fact, this transition is already happening. HP embraced this future in 2013 when they created "Instant Ink" - a service enabling connected printers to automatically order their own ink when supplies run low. The premise is simple: pay monthly, based on how many pages you print and delegate ink purchasing to the machine. The printer makes sure ink is delivered to your door before you run out and the entire transaction takes place without any human intervention. HP is, in effect, manufacturing its own repurchasing customers - smart printers. Diana Sroka, Global Head of HP Consumer Print Services describes it this way:

> *"Instant Ink was really invented to solve customer pain points around printing. It is an important utility for most people, but what we hear from customers are usually two pain points. They say ink is expensive and they run out at the most inconvenient times. So, HP developed a service crafted to remove those challenges and allow for customers to really enjoy the outcome of printing, for example family photos, recipes, and their children's' art. What we do is make a promise to the customers: never run out of ink and save. And I'd say 11 million subscribers and nine years later, that promise continues to resonate with customers around the globe."*[6]

Others are headed down this path. Amazon's Alexa virtual assistant is now compatible with at least 140,000 different products including cars, bicycles, and toilets.[7] Many of these devices embed an Alexa service capable of buying on your behalf as an Amazon Prime member for example:

- A Keurig capsule coffee machine that can reorder its own capsules[8]
- An Oral-B smart toothbrush that automatically orders brush heads[9]
- A Winmax air purifier that can order its own replacement filters[10]

Other more advanced Machine Customer enabling models are emerging.

- Walmart started using a chatbot from software provider Pactum, to negotiate directly with suppliers.[11]
- A consumer-support company called DoNotPay created GPT-powered chatbots that emulate human customers on customer service lines to help their clients handle such chores as renegotiating bills and cancelling.[12]
- Intermediaries are starting to offer the trust and payment infrastructures needed for machine customers and sellers to trade at scale such as Mastercard, Visa, PayPal and Stripe[13]

These early examples of generative AI (or GenAI) are enabling machine customers to access companies via human-style conversation and learn from market activity data more quickly. We think that in many cases, these Machine Customers will evolve to become more effective than human customers at buying products or services. Machines will not be swayed by the emotion that often influences human purchasing decisions. They will rely on calculation, rules, and logic. They will not be swayed by advertisements that would tug at your heartstrings, nor wooed by being taken to a nice dinner or given tickets to a basketball game.

The benefits are numerous:

Machine Customers can process large amounts of data from a wide variety of inputs using AI. They will carefully collect and weigh the data to make an informed decision based on patterns, logic, and rules - acting as educated guess-making algorithms that will enhance the human customer experience even as it advances their own capabilities as customers.

Machine Customers don't need to be delighted. Resources normally spent on exceeding human customer needs can be diverted to making the machine transactions seamless.

Machines may be more willing to lock into a supplier. When the sales and fulfillment process work seamlessly and meets the requirements of the service level agreement, human customers will appreciate the convenience.

Machines may be more adaptable to human requirements and changes in product specification. Humans change their mind frequently. Machines will have no problem keeping up.

Machines will be more reliable and consistent. You can ask a machine to keep reordering toothpaste, and it will never forget to repeat the order. Complex corporate chores will be handled with similar reliability: one B2B startup, VIC.ai, is working now to automate accounts payable.[14]

Machines will minimize waste by ordering precisely what the organization needs at the right time - promoting more sustainable productivity. This could take the form of smaller sizes or just-in-time replenishment, depending on availability from the provider. Upshop, an inventory optimization system for the grocery industry, offers automatic re-ordering and replenishment that can adjust for real-time sales, inventory levels, safety stock, promotions and display quantities.[15]

Machines will be better able to find substitutable products and request products and services before they are needed. They may even recommend value-added products that might be more expensive in the short run, but cheaper in the long run (such as a service contract or warranty).

Economies and societies need growth – it is who we are. Business always does better when it can pursue market growth rather than market share. From the early twentieth century heyday of the oil giants to the ascendency of the tech titans of today, total market growth has created great wealth and progress. For those who pursue it, the emergence of Machine Customers will spur another era of economic opportunity and prosperity. Even if Machine Customers can't be delighted, we think their human owners will be. Machine Customers will be part of the next giant leap in capitalism.

How Will Machine Customers Work?

Today's digital channels will evolve to become tomorrow's Machine Customers. We believe the digital giant corporations, with their vast resources and desire for new growth, will inevitably drive this shift. Machine Customers will rely on a complex web of capabilities, technology, data, and ecosystems, much of which already exists.

We believe the core capabilities of Machine Customers will follow a four-step process analogous to military strategist John Boyd's OODA model – Observe, Orient, Decide, Act[16]. Machine customers will:

- **Understand** the needs and constraints that would drive the criteria for the selection of a solution as directed by a human customer

- **Identify options** and relevant information related to those needs and constraints
- **Decide** from among a set of choices, according to the parameters established by the human customer
- **Buy** the product or service on behalf of the human customer

The primary technologies and related capabilities that make Machine Customers possible are already in use today:

The Internet of Things. If a physical machine is to become a customer, it must be connected to a broader set of systems. Today, almost any product can be designed to communicate. Whether it's a fitness tracker or an oven, many devices have capabilities that work best when connected to the internet.

Edge and Cloud Computing. Edge computing is essential because it pushes as much computing power closer to the device as possible – in our case, the Machine Customer. Of course, the complement to the edge is the core or what is broadly known as Cloud Computing. That is, the upstream system or systems that guide what is passed down to the systems at the edge, often serving as the centralized storage, processing, and archiving "back end." This computing network gives life to Machine Customers, transforming them from isolated devices to part of a vast intelligence network with its own set of possibilities.

Artificial Intelligence. If we compare a Machine Customer to a human, the device would be the body, while edge and cloud computing would represent the nervous system. These are both essential, but they can't go far without the artificial intelligence that would constitute the brain. The rapidly advancing capabilities of AI (generative and agentic) fed by robust data, are the magic wand that will transform Pinocchio from a marionette into a real boy.

Enterprise Data Management. The Machine Customer is fundamentally a data-driven entity. Unlike human customers, machines are entirely reliant on data to operate, decide, and learn. Their effectiveness, and the trust humans place in them, is directly proportional to the quality, accessibility, and breadth of the data they can gather and interpret. This urgent demand for data to support AI initiatives, including those powering Machine Customers, requires evolutionary changes to traditional data management.

Platforms and Ecosystems. Platforms and ecosystems will matter because scale and openness are vital to Machine Customers. These platforms will range from high-level platforms that facilitate a market (think Amazon) to low-level platforms that provide a collection of business and technology capabilities that other products or services use to deliver their business capabilities (for example, Airbus' Skywise Platform[17]).

Trust. In almost every interview we conducted for this book, executives saw trust as a major impediment to building a world of Machine Customers. First, they said humans will have to trust machines to perform tasks they usually did. Second, organizations will need to trust the Machine Customers of other organizations and be willing to share data for the benefit of all players in a Machine Customer ecosystem. Third, organizations will need to trust that Machine Customers are the right strategic choice for the growth of the business. Trust will need to be built over time, starting with the lowest-risk tasks and situations, and working up from there. As the need to serve Machine Customers grows, businesses will take the risks required to create this trust environment.

Machine Customers are coming because it is almost human destiny to create them. From using fire to cook meat to reduce chewing time, to the invention of the dishwasher, we have always used technology to conquer current work. We do it because it liberates time for us to do other things. The technology that allows us to create Machine Customers is now reaching a critical tipping point as the companies we call digital giants forge the infrastructure required.

Tech giants are those very large companies born in the internet era that use digital as a core competency for competitive advantage in non-tech traditional industries such as retail, media, or automotive. Examples include Amazon, Google, Netflix, Tesla, Tencent and Alibaba. Many of them are already competing for a share of the system that will power Machine Customers.

How Will Machine Customers Change Your Business?

Machine Customers will often be superior to human customers. Those companies that get it right can manufacture customers - at scale. Companies that learn quickly how to create these customers and others that learn how to sell to them will win big. Others could find their traditional human buyers gradually disappearing - without even understanding why.

But nothing in business comes without risk, and very few things come without a price. Creating, serving, or managing Machine Customers will not be free and will introduce many new complications.

The biggest changes will be in Marketing and Sales. Today a great deal of marketing and selling exploits emotion – both in B2C and B2B. Machines don't have emotion - they will behave dispassionately and rationally based on their programming. If you are a leader in a business that relies on the mostly rational behavior of human customers today, then moving to a machine as a customer should not be an issue. But suppose emotion does play a role in influencing the purchasing decision (e.g., luxury goods, indulgence, B2B wining and dining). In that case, selling to a machine will present a host of challenges, especially when it substitutes the manipulable irrationality of the human customer. All business functions will have to adapt to a world of Machine Customers - from Finance (e.g., cash flow visibility) to Legal (machines as corporate persons), to IT (cybersecurity). And with so many departments and specialists in flux, the HR leader will also be busy, trying to find people with the right skillsets and sensibilities.

Boards of directors must be involved because of the scope and scale of the Machine Customer challenge. Going after whole new markets will always be something a CEO discusses with the board. It won't be restricted to the C-suite. Strategy, opportunity, threat assessment, corporate values, and company purpose may change. As Machine Customers emerge in markets, the issues thrown up will be material to the company's future and potential investors.

Smart companies will break their capabilities into two realms: Techs and Strats. The Techs will mastermind the platform needed to engage Machine Customers, and the Strats will focus on strategy – crafting the algorithms required to sell to a Machine Customer. Organizations like HP, Allianz, Cummins, Unilever, and Salesforce are already hiring teams as they develop various aspects of Machine Customer functionality. Major credit card companies hope to be trust brokers that can ensure the security and validity of Machine Customer transactions.

The biggest change we see is the shift from B2B to B2T (things) or B2A (AI assistants). All the models of the internet era so far have exploited value flows between just two entity types: human individuals (consumers) and human organizations (businesses). We have shorthand names for the categories – B2C, B2B, C2C, etc. What happens when we add new machine actors into markets – things and AI assistants? The result will multiply the business model map – B2T, B2A, C2T, B2A, and so on. Once machines become economic actors, we expand the number of business model possibilities.

Aaron Rajan, Global VP of Consumer Digital Experience at Unilever, shared this about B2B selling to wholesalers and retailers:

> *"We are selling to an algorithm. That is entirely the case. There's a buying algorithm that has rules. There's an exchange of information. There's sometimes an arbitration process if there's a dispute over whether goods arrived or not, which will be facilitated through an algorithm. So, I think it is quite interesting. It's a bit of an arms race to kind of call it, as to how you keep ahead. Especially with something like dispute management, did the product arrive on time? How do the machines resolve disputes? So, we're very much engaged in AI right now. It's very real."*

The rise of the custobots will change the fortunes of organizations that take this trend seriously. They will increase total sales in many markets by being more efficient, timely, and consistent purchasers. They may reduce waste by making smarter package size choices or lower energy use (and shipping costs) by favoring near shore suppliers. Or they may recommend value-added products that might be more expensive in the short run but cheaper in the long run. Machine Customers also won't forget to buy things, and their vigilance will add billions of new sales opportunities.

However, Machine Customers will also create two major new worries for your company. First, many businesses will need to double up their customer operations infrastructure to provide for both Machine Customer AND human customer needs. Depending on the industry, they may have to maintain two sets of overheads for a number of years – not just in channels but in product design, price points, and other factors. Second, when machines become customers, they act much faster than humans. Often their needs can only be met by other machines. Machine-to-machine marketplaces can be incredibly efficient, but they can sometimes go horribly wrong at terrifying speed. Remember the "flash crashes" in the U.S. and other financial markets? Those were the result of "robo-trading" run amok. The same thing could happen in any electronic market.

We have also identified several scenarios where Machine Customers go "bad," meaning that businesses could be damaged or even destroyed through the intended or unintended consequences of using Machine Customers. And, in the process, people could be hurt. The good news is that we already have ways to mitigate these risks, using many of the same processes and technologies that secure our digital infrastructure today.

The temptation to drag your heels, rely on dwindling human customer markets and fail to redevelop critical competencies will be high. But don't let yourself be lulled into thinking Machine Customers won't affect you. There will be winners and losers. In many sectors, Machine Customers will have significant downsides for many companies, as their ruthless efficiency winnows markets down to last-player-standing.

What Are the Consequences of a Machine Customers Economy?

On the positive side, Machine Customers will spur economic growth in the countries that embrace them. Machine Customers will reduce waste, as they can be relied on to buy only what they need. The arrival of Machine Customers will lead to new areas of economic growth that policymakers will need to harness and help to develop actively. Often, the conversation will be about existing kinds of machines that will gradually become customers. But some of the Machine Customer markets that become apparent early on will be entirely new kinds of machines that have not existed before. As new forms of intelligent digital products emerge, they will be customers from the get-go. New adjacencies and product opportunities will arise as Machine Customers hunt for feature and function variants that don't yet exist and expose market gaps that humans had previously overlooked.

Machine Customers will progressively take on the drudgery of customer work done by humans. This could reduce buying cycle times, remove unnecessary repeated interaction steps, reminders, missed sales meetings, and many other sources of inefficiency. It could also enable faster company formation through lower barriers to entry and the possibility of testing start-up ideas in real markets while burning through lower amounts of venture capital. As Machine Customers proliferate, the effect should add predictability and coherence, leading to more manageable volatility. Overall, then, as market efficiency improves, the economy should grow stronger.

Governments will like Machine Customers too, because economic activity will be easier to trace – and tax. "Who" bought what, when, where, and why will be recorded by Machine Customers. In that sense, Machine Customers might be a boon to governments, especially in countries where the informal economy has traditionally been large.

The losers will be those companies whose sales are artificially bloated because they prey on human frailties. Especially in the early stages of the Machine Customer revolution, those markets could shrink. Companies that profit by selling us things we may not need, encourage our FOMO (Fear of Missing Out), or trick us with Buy One Get One Free (BOGOF) offers, deceptive pricing, and scams hidden in small print will all have to find ways to add more real value. Add into the mix social media and the enormous amounts of misinformation it can create, and the situation only looks worse.

Also on the negative side: Machine Customers will expose artificially distinguished products, leading to rapid commoditization - a frightening thought for companies that compete in crowded fields oversaturated with choices (does anybody really want 200+ kinds of potato chips?). It's also not good for companies that rely on place and time convenience to win a higher price (such as hotels, airlines, and convenience stores). Machine Customers will be forever watchful and patient, as happy to buy at two in the morning as they are at three in the afternoon, in a way that most humans aren't. The ruthless efficiency of Machine Customers – particularly if they sweep into a market quickly – might lead to the failure of individual companies that cannot adapt fast enough.

While Machine Customers will and must operate on behalf of their end-users, their creation and maintenance won't be conducted by those users themselves. You didn't program the Instant Ink service in your printer, and you won't be the creator of the tire buying capability of your autonomous car. The providers of Machine Customer capability will be companies seeking the kind of scale effects we have seen in other online platforms like mobile operating systems, digital commerce platforms, entertainment platforms, and social media platforms. Their success could lead to various kinds of abuses, most of which can be overcome with strong, insightful regulation – hopefully sooner rather than later, after a market crisis.

What Will Be the Impact of Machine Customers on People?

Machine Customers will offer many good things for people who choose to join that ecosystem. Whether you make them, sell them, or serve them, we believe they can add value to your personal and work life. Ultimately, they offer a level of automation and convenience that will make your life easier if you let them.

There is a bright side to the arrival of Machine Customers. For example, Machine Customers will take away some of the drudgery in our lives. Tasks like grocery shopping, sweeping the floor, taking out the garbage, cleaning the windows, and so on will be taken off your plate, one by one. In fact, it's already happening, as we show throughout this book.

Machine Customers will improve your confidence in buying decisions. When we buy products or services, there is often a nagging voice in our heads that says, "you could have done better" -- a lower price, better terms, or a special feature we forgot to order. Machine Customers won't have that problem. They will patiently collect and process far more information than we can as humans. And, with the proper guidance from humans, they will identify, recommend, and make the best possible decision – at least according to the criteria of their algorithm.

Machine Customers will protect you from your weaknesses. We all do things that, in a perfectly rational world, we should not be doing. Like having that slice of cake when you are on a diet, skipping a workout, or buying that piece of clothing when you have three other things just like it. Marketers entice us to do things maybe we shouldn't do. It's part of the process, and we all succumb to their tactics. Just as Machine Customers can increase your confidence when you buy things, they may also be equally good at making sure you <u>don't</u> buy things that don't align with your goals – at home and work.

However, there is also a dark side to Machine Customers - their advantages will come at a price. In the case of Machine Customers, they will take away things we miss from our past lives as human customers. We will feel some of the effects immediately, while others will take time to seep in. The most important is the loss of control. When we delegate a task to someone else, we must let go of the direct responsibility for that task and trust that someone – or now, that some*thing* – to get it done. When delegating to Machine Customers, we might not fully understand how the machine makes decisions or have trouble explaining why a machine made the decision it did. We might doubt the ability of the machine to execute a task. And if they do a good job, we might feel threatened when the Machine Customer does our job better than we could.

We will lose some degree of choice when we delegate to Machine Customers. When trying to find the best choices for your needs, their priorities will likely approximate what you want. That's because Machine Customers have no emotions, so they will rely on logic and rules. We might also lose some of the joy of shopping, both in the joy of the hunt and in the social pleasure of rubbing elbows with many other people. You might run into a friend at the mall. Your custobot won't.

Machine Customers could also cost you your job, particularly if you work in marketing or sales. We have no reason to believe that the arrival of Machine Customers will destroy more jobs than it creates, but like every other wave of technology innovation, things will change. And, just as businesses will win or lose in a world of Machine Customers, so the same will happen with people's careers.

On the other hand, there will be a large variety of roles to play in the Machine Customer transition: change leaders of all kinds and at all levels, entrepreneurs, and new kinds of specialists. Business transformations of scale consume hundreds of thousands of graduate-level careers. Just imagine how many people have been involved in transitioning companies to Digital Business in the last few years. The arrival of Machine Customers will impact different industries at different times. Significant market effects will start arising early on, but we think the change wave will take at least twenty years or more – just as e-commerce did. So, for those who want to get involved, there will be many career opportunities.

We expect a significant role for government and regulatory organizations. Machine Customer market evolution cannot be left to so-called 'free markets' alone. That never happens in practice. From the evolution of mobile telephony to the advent of unmanned aerial vehicle (UAV) drones, governments have always had to step in to set rules, guidelines, and standards. Without them, market evolution can take harmful directions or end in exploitative monopoly, or just massively underperform. So, the advent of Machine Customers will inevitably create many roles for specialists who learn how to develop standards, rules and controls to open Machine Customer markets, and ensure they operate smoothly. There will also be roles for people who police standards and investigate breaches or unexpected situations.

Will The Machine Customer World Create an Opportunity for Me?

That is the big question, right? That is what you need to know about all of this. The answer is yes if you are prepared for what is coming and open to change. The Machine Customer world will not pop up overnight. There will be plenty of time to prepare and adjust. We think that no matter what you do today, there will be a different version of it in future. The trick will be to see it and to seize it. Look no further than the changes brought about by digital commerce, the changes underway with autonomous machines, and the current shift to clean energy. There will be a myriad of new opportunities for you, but you may have to realize, sooner rather than later, that staying put will not be the best strategy.

You might tell yourself that you love the human aspects of customer-related work and that machines just are not the same. That is a natural reaction – but it could cost you. Instead, you might ask how much real customer contact is in your job today. A lot of it is an illusion. We already work at arm's length in digital and e-commerce remote businesses, often with low direct human customer contact.

It's early days for this of course, but it will only get more complex. You will serve yourself well by understanding what Machine Customers mean for your personal and professional life and following the best career path.

Where Will Machine Customers Lead Us?

We wrote this book to lift our heads up from the day-to-day business of our lives and imagine the possibilities of a future where machines, in the form of customers, would be a force for positive change.

Machine Customers Are a Powerful Trend, So Pass It On!

When any new megatrend arrives – from sustainability to blockchain – the best way to get ahead is to become known as the wise one and the clarifier. At the highest level – you should learn how to introduce and teach Machine Customers to others. Why teach? Because when a new wave hits, it confuses. The people who make the most progress are the ones who can peer through the fog and help others make sense of what is going on. When you teach, you also learn.

Regardless of your leadership role, what will most of your peers do when the concept of a Machine Customer starts to become real in your sector? Probably nothing. They will tend to avoid it and stick to what they know. To be a true leader, you must engage with the ideas, learn about them and get yourself into a position to confidently show the way to others. It might sound easy – but going first is often a little uncomfortable. You must resist the need to be an expert before saying anything and start finding ways to explain.

We Must Trust The Machine Customers and Let Go to Grow

We won't deny it, when machines take on the work of humans, something is lost in the process. We risk becoming less human. On the other hand, some say that being freed from tedious or repetitive tasks frees humans to pursue higher-value opportunities. As Jeff Bezos, Chairman of Amazon, once said: "I predict that, because of artificial intelligence and its ability to automate certain tasks that in the past were impossible to automate, not only will we have a much wealthier civilization, but the quality of work will go up very significantly and a higher fraction of people will have callings and careers relative to today."[18]

We ask the question: what will it take for humans to embrace Machine Customers? Based on our research, it comes to three things: trusting the technology, trusting the ethics of Machine Customers, and trusting yourself.

Trusting the technology in context. Humans must reach a level of trust with Machine Customers not only that their underlying technology is reliable, but that they will make contextually aware decisions. Humans must trust that those decisions will provide the best outcome for their specific needs. Custobots must be as trustworthy as humans, not only in executing on explicit instructions, but also understanding implicit instructions and acting in the human's best interest. Trust in technology to do the right thing is what is needed.

Trusting the Ethics and Privacy of Machine Customers. Suppose the decisions of Machine Customers are grounded in some algorithmic pattern. In that case, businesses, humans, and other machines have an opportunity to predict the Machine Customers' behaviors and take advantage, both ethically and unethically. In some cases, bad human actors might use Machine Customers to cause damage. Negligent operators could also be harmful: Developers are struggling to find ways to limit the "hallucinations" fabrications and logical errors to which generative AI is currently prone.[19] Regardless of the degree of risk, a strong moral and ethical compass is needed to use Machine Customers. We need digital ethics.

Trusting Yourself and what it means to be human. Machines are good. Humans are better. Machine Customers are just one in a lengthy list of automation actions that have made life progressively easier for us as humans. The rise of conversational platforms and virtual assistants is the clearest indicator of what is possible in a world of Machine Customers. Systems capable of learning more about our purchase and consumption behaviors and recommending the next best actions to take the drudgery out of our lives are quite appealing. But if you are like most of us, delegating the purchasing of laundry detergent does not challenge what it means to be human. Human intelligence, creativity, and initiative will always win out over machine intelligence. Because machines today are not creative, they can't take the initiative unless programmed, and their intelligence has real limits. We should take confidence in this: we will always have the off switch.

We Must Free the Machine Customers to Expand Horizons for Us

The authors are both unashamed sci-fi lovers. Don's array of Star Trek collectibles is a wonder to behold. In our careers as research analysts, we have found that the long-range futures imagined and explored by sci-fi authors illuminate what annoyingly plausible technology disruptions might be not so far away. There is no shortage of entertainment content today - from the resurgence of the Star Trek franchise to original series like Amazon Studios' "The Expanse" and the imagining of Isaac Asimov's work in Apple TV's "Foundation". All involve the human need for exploration and expanding the possibilities of what humans can achieve. For some intrepid billionaires this is about space. And machines will play a big role in that future.

An almost entirely machine-to-machine space economy is not hard to imagine. Satellites need fixing, upgrading, and deorbiting. Space junk needs collecting. Companies are setting themselves up to provide services to satellites. Of course, initially, this will all be organized and negotiated down here on earth between humans. But soon enough, that will become a problem. We are too far from the action to manage it well. We will need specialized servicing spacecraft to negotiate, price, transact and serve satellite customers more directly.

So, Machine Customers will be a critical component that enables our economy to continue growing off-earth. What is that realm worth? Morgan Stanley has conservatively estimated in 2022 that the space economy will be worth around a trillion dollars a year by 2040[20].

Machine Customers Must Have Rights and Responsibilities to Thrive

The idea of personhood or citizenship for machines (and now AI) is an ongoing debate with no immediate conclusion. Do machines have rights? Can they be held accountable for their actions? Do machines have feelings? Science fiction authors like Isaac Asimov and others have explored these questions' implications for decades. Now that we are on the cusp of technological advances, these questions are becoming more and more real. For this approach to become normalized, we will need to frame the rights and responsibilities of Machine Customers in a way that will create and defend trust. To help with that, we have formulated a draft model set of rules inspired by the famous Asimov Laws of Robotics.

How Do You Prepare?

Entering the world of Machine Customers may seem daunting. It doesn't have to be. In the dozens of interviews with business, technology, and academic leaders we conducted for this book, most experts told us that taking small, but significant steps is what you need to get started. This includes seeing the possibilities in your ecosystem, picking a direction, and doing something to get started.

When it comes to seeing the possibilities of Machine Customers, the catalysts can come from a variety of directions – an article, a short video on your phone, a discussion with colleagues, or maybe even a book. More likely, it starts with a question such as "What is a Machine Customer?" or "Is this a threat or an opportunity for us?"

Each chapter of the book provides answers to a big question about machine customers and their implications. At the end of every chapter, we also provide a short list of actions you can take to apply what you have learned. To get you started, here are the highest-level actions to consider.

Create A Machine Customer Investigation Team.

This exploratory team comprises senior representatives from strategy, IT, product development, sales, and marketing. You want a team of people that share your interest in the possibilities of Machine Customers, but also bring their own experience and biases to the party. You want a "team of rivals" that can challenge each other yet are committed to the longer-term vision. This team would be tasked with the actions that follow.

Formulate Several Scenarios That Explore the Market Opportunities.

What IoT-enabled products might arise in the situations and activities where customers use your products and services today? Who might create and control those Machine Customers? How would Machine Customers change your addressable market – or even what business you are in? The idea is to embed

Machine Customers as a "big rock" in your long-term strategy.

Test Products or Services Capable of Being Upgraded for Custobots.

Use a capability and cognitively diverse team of technology architects, product leaders, engineers, data scientists, economists, linguists, psychologists, R&D leaders, and business decision makers that will determine assessment requirements, profiles, recommendations, and action plans for engaging machines as customers.

Be Mindful of The Very Real Barriers.

The complexity involved in developing an intelligent custobot that can learn the depth and breadth of knowledge and preference tradeoffs required to act on behalf of a human customer in a variety of situations is complex. Some humans may initially be uneasy about delegating purchasing functions to machines. Most importantly, the technology must work. Consider what ethical standards, legal issues, and risk mitigation are needed to operate in a world of machines as customers.

Look Beyond Earth, to Space.

Yes, this seems a bit far off, but not so very far. As private enterprises and government space agencies evolve the space transportation, exploitation, and exploration industries - opportunities will be available for Machine Customers, for those with foresight and courage.

Machine Customers may be the last and largest emerging market of customers on Earth and may well extend beyond it. Digitally savvy organizations will realize their best customers may not be human. Just as the world has adapted to mobile devices and now artificial intelligence, so we will adapt to Machine Customers.

Billions of Machine Customers are headed your way. Ready or not.

Someone will make them.

Someone will service them.

Someone will sell to them.

Someone will buy from them.

Will it be you?

Chapter 2

Meet Your New Machine Customers

Welcome to a world where your best customers may not be human

What on Earth is a Machine Customer?

Today, virtual assistants (like chatbots and AI agents) can perform a growing array of tasks for us. Every day new skills, connections, and functions are becoming available to make our lives easier at work and at home. Machine Customers are a subset of this world of virtual assistants and internet-connected physical devices - focused on shopping, buying, and consuming, just like us.

A Machine Customer is a digital agent that acts on our behalf - either a virtual assistant like Siri or Alexa, or a physical object connected to the internet, like your car, your washing machine, or a factory robot. It will act as a concierge who can take care of the tedious side of shopping and purchasing for us. Eventually these machines will evolve to shop for themselves as well.

More precisely, we define a Machine Customer as: *A non-human economic actor that obtains goods or services in exchange for payment.*

We know the term Machine Customer can sound a bit like "MBA-speak", so let us introduce a more down-to-earth alternative: custobot. We will use these two terms interchangeably throughout the book.

Two Kinds of Machine Customers: Physical and Virtual

The first type of Machine Customer is a physical device that connects to the Internet or a local network. These devices will be built with a variety of sensors that collect external information like motion, sound, temperature, and light. They will also generate information specific to the performance of the device itself, like efficiency and power usage, and whether the device is being used as designed. The value of this type of machine acting as a customer is in the connection between the physical and virtual worlds. The internet-based system collects this internal and external data and turns it into insight and recommendations. Embryonic Machine Customers can be found in cars like Tesla, smart homes using Google's Nest home automation hub, and trains from Toshiba. In all these cases, the manufacturer has a proprietary system that allows the human owner to issue commands and communicate with the machine. Gartner estimates that there are at least three billion installed IoT-enabled B2B devices today with the capability to be customers--we expect that number to rise to 7 billion by 2030.[21] The number of consumer devices and virtual AI assistants is expected to be 2-3 times that number.

The second type of Machine Customer is a general purpose, virtual system. These are virtual personal assistants (VPA) and buying algorithms. You might also think of them as AI assistants using Generative AI or Agentic AI techniques that take the form of chatbots. Here are some key definitions that will be important throughout the book:

Generative AI refers to artificial intelligence techniques that learn a representation of artifacts from data and use it to generate brand-new, completely original artifacts at scale, that preserve a likeness to the original data (including images, video, music, speech and text).

Agentic AI is an approach to building AI solutions based on the use of one or multiple software entities that are classified, completely or at least partially, as AI.[22]

AI Agents are autonomous or semiautonomous software entities that use AI techniques to perceive, make decisions, take actions, and achieve goals in their digital or physical environments.[23]

Examples include Tencent's Yuanbo, ByteDance's Doubao, Deepseek, Apple's Siri, Amazon Alexa+, Anthropic's Claude, Google Gemini, Perplexity and Microsoft Copilot.[24] The B2C systems can operate through a variety of voice interfaces without being limited to one type of device. Amazon's Alexa has expanded from its consumer base into business. We also see the emergence of business-specific virtual assistants like Salesforce's AgentForce. As their numbers and level of collective experience grows, we expect they will become a powerful foundation for what it takes to be a successful Machine Customer.

Soon, we expect that both types of Machine Customers will be so ubiquitous in our lives that the lines will blur between the two. For example, according to Amazon, Amazon's Alexa virtual assistant is now compatible with at least 140,000 different devices.[25] Those range across dozens of different product categories, from cars to bicycles to toilets. Some manufacturers, such as Ford, are forgoing development of their own assistants to incorporate systems from companies such as Google.[26]

Wait – Doesn't a Customer Have to be a Human?

That used to be the case, but not anymore. To help show you that, we'll first break down what it means to be a human customer.

A customer can be a single person, a family, a business, a corporation, or even a country.

A customer has money to spend and could be buying for themselves, buying for someone else, or buying for an organization.

A customer has goals. We have needs to be met. We might even be on a mission. As customers, we are seeking to solve a problem by buying something. Our approach might be dominated by pure emotion or governed by a strict process and anything in between. Shopping for and buying things is an ancient human activity. It is also a social activity where buyers and sellers connect.

Being a customer is a big job when you break down all that's involved. We:

Process messages from businesses trying to sell to us. Whether we buy or not, we filter hundreds of television commercials, billboards, email, text messages, and ads in social media every day. We pay attention to some (if they are relevant) and ignore the rest.

Search for information. We spend a lot of time deciding what to buy. Today, this often begins with a search on Google or Amazon, but may also include consulting user reviews, and even (gasp) talking to friends, family, or co-workers.

Negotiate for the best deal. We also spend a lot of time negotiating, particularly for big ticket items. We love (or hate) the thrill (or chore) of negotiating – with a car dealer, at an open market, or with another company.

Buy goods or services. We hand over cash, swipe a card, or press a button, and become the proud purchaser of something.

Request service when something goes wrong. When an item does not arrive, a charge appears that we did not make, service is delivered poorly, or a machine breaks down, we contact customer service to pressure them to make things right.

Tell others about our experiences. If a purchase goes well, we tell our friends, and now with social media, millions of strangers. And if something goes badly, we may tell even more people about it.

Whether you enjoy these tasks or hate them, they currently take a lot of time. But not for long. We believe that soon machines will be able to take over some of these tasks, and eventually almost all of them.

This is not to say that humans will be irrelevant. Humans will continue to make the rules and if need be, humans will be able to flip the off switch. But our belief is that there are enough advantages to this next phase of automation that you will probably let the machines take care of most of these tasks.

The Internet of Things and AI have Changed What Customer Means

With the rise of the Internet of Things (IoT) and ever-present artificial intelligence, machines have been quietly taking on more of these five core customer tasks:

Figure 1 - Some of the Activities that a Machine Customer can Undertake

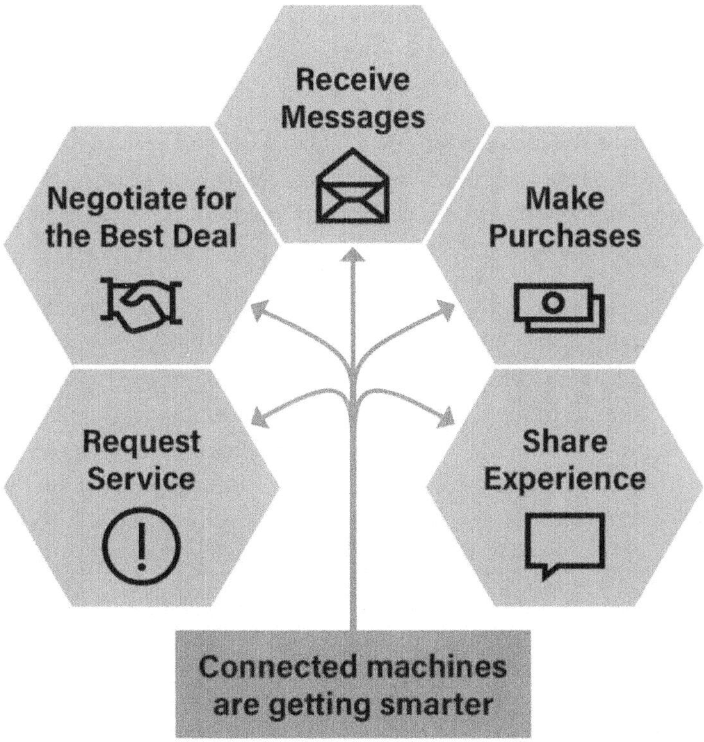

As illustrated in Figure 1, today machines can:

Request service when something goes wrong. Tesla vehicles can now diagnose themselves and pre-order parts for service[27] The Apple Watch can call emergency services if you take a hard fall (from a bicycling accident, for example) and don't seem to move.[28]

Receive messages from sellers. Carvana, the online auto trader, receives thousands of messages from people asking for quotes, trying to sell their cars. Carvana built a car-buying offer algorithm that performs similarly to human bidders at car auctions. The system combs through public databases that record insurance, registration, mileage, accidents, and more before it issues that initial quote. The company trusts its algorithms to act as a buyer – offering a guaranteed sum of money and cutting a pre-printed check, instead of haggling in the driveway.[29]

Buy goods or services. Amazon Dash Replenishment services use a smart shelf, at home or at the office, which can trigger a re-order of a consumable product once a certain usage threshold is reached.[30]

Tell others about their experiences. Today, bots can post product reviews on websites. However, bot reviews are considered fake by organizations and are seen as a nuisance. This will need to be addressed as the use of Machine Customers scales.

Negotiate for the best deal. Most of us are familiar with the concept of "Robo-trading" - computers that use sophisticated algorithms to buy and sell shares on the stock market with other computers at lightning speeds. Closer to home, Harvest, a U.S. based fintech, uses AI to help consumers negotiate bank fees and get refunds automatically.[31] Walmart has also deployed AI-powered negotiation software with a text-based interface (i.e., a chatbot) to connect with suppliers of "goods not for resale" — categories such as fleet services, carts, and other equipment used by its retail stores.[32] AI Haggler is a web-based application that utilizes AI to call hotels and negotiate discounts for last-minute bookings. Users can input hotel phone numbers, select their preferred language, and set negotiation preferences. The AI agent then contacts the hotels to inquire about room availability and negotiate better rates. After the calls, users receive detailed reports with the results, enabling them to make informed booking decisions.[33]

In addition to this list, we see two other capabilities emerging in Machine Customers:

Anticipating customers' needs. This happens when a system provides a recommendation or "next best action" to a customer. Most of us are familiar with the suggestions we get from digital commerce platforms ("people who bought this, also bought that") and increasingly sophisticated forms of product curation. For example, StitchFix is an online retailer that uses AI and human stylists to customize a selection of clothing and accessories based on your purchases, preferences, and behavior of those like you.

Fulfilling their own needs. We see this today with industrial equipment requesting its own service. Honeywell offers an IoT-based predictive maintenance solution for distribution centers, allowing DC operators to address critical issues before they happen, minimizing unplanned downtime.[34]

Machine Customers Have Been Around Longer Than You Think

Machines have been buying things on our behalf for over 60 years, but their work has been largely invisible.

From LEO To Alexa: Custobots' First 60 Years of Purchasing

Like the Starbucks of its day, J. Lyons and Co. was one of the UK's leading catering and food manufacturing companies in the first half of the 20th century. In 1947, it sent two of its senior managers, Oliver Standingford and Raymond Thompson, to the US to look at new business methods developed during World War II. During the visit, they met Herman Goldstine, one of the original developers of ENIAC, the first general-purpose electronic computer. Standingford and Thompson saw the potential of repurposing military computers to help solve the problem of administering a major business enterprise. They recommended that Lyons should acquire or build a computer to meet their business needs. In 1951, the Lyons machine was christened Lyons Electronic Office, or LEO.[35] One of its first tasks was managing all the daily orders, which were phoned in every afternoon by the tea shops across the country and used to calculate the overnight production and delivery. LEO was the first instance of an integrated management information system that facilitated purchasing and the first step in a long evolution toward Machine Customers (Table 1).

Table 1 – The First Century of Machine Customers

Decade	Machine Customer Techno-evolutionary Milestones
1950s	• LEO, the world's first order collating computer
1960s	• Electronic Document Interchange (EDI) standard for orders
1970s	• UPC code standards and the first use of barcodes
1980s	• Rapid growth of computerized delivery logistics- DHL, FedEx, UPS • Teletext - Mintel, Prestel, and other online shopping precursors
1990s	• The emergence of web-based electronic commerce • Amazon's collaborative filtering
2000s	• Search engines • Mobile computing • Cloud based open API e-commerce platforms
2010s	• eBay sniping software • Amazon Alexa and Apple Siri • IoT connected products and the smart home
2020s	• Bound and adaptable Machine Customers such as o HP's Instant Ink and o Amazon dash replenishment services o Shopify's Shop.AI o The emergence of AI Agents acting as Machine Customers: OpenAI, Google, Perplexity
2030s	• The emergence of autonomous Machine Customers at scale • Mobile robo-taxis and drones that earn their own money and support themselves
2040s	• Machine Customers gain limited legal personhood • Machine Customers become a key part of the expanding space economy
2050s	• Self-replicating Machine Customers?

In the 1960s, the U.S. transportation industry developed EDI (Electronic Data Interchange) to standardize electronic communications between customers and vendors. EDI offered companies many advantages – it reduced manual data entry errors, streamlined transaction processing, increased productivity without increasing staff, and made it easier and more cost-effective to do business with other companies. The U.S. grocery and automotive industries were the first to embrace the standard. Today all major industries employ EDI, and many must comply with its standards that govern the flow of electronic documents between companies, streamlining and enhancing the supply chain.

The next big advance toward the world of custobots came with the advent of the world wide web in the 1990s. In 1995, a 31-year-old ex-investment banker named Jeff Bezos started selling books out of a rented garage in Bellingham, Washington on something called a web page, and a new purchasing channel was born. Soon, as secure payment methods and shipping speeds improved consumers and businesses overcame their initial fears of buying things over the internet.

Flash forward 28 years, and we can ask Alexa (powered by a ChatGPT Skill[36]) to help select a toilet tissue brand, then place our selection in the shopping basket, complete the purchase and arrange for it to show up on our doorstep the next day – a transaction barely conceivable even 10 years ago.

Global eCommerce sales are projected to reach 20% of all retail sales worldwide in 2025.[37] This figure is expected to rise to 22.5% by 2028.[38] The 2020-2022 global COVID-19 pandemic undoubtedly accelerated this trend.

Computer Programmed Buying and Selling

Custobots might still seem like science fiction on Main Street, but they are practically middle aged on Wall Street. Algorithms have been deeply involved in buying and selling securities for the last 40 years, but the public became aware of just how deeply only after the publication of the 2014 bestseller FlashBoys[39] in which author Michael Lewis told the story of how high-frequency trading between computers was transforming finance, as algorithms bought and sold securities in nanoseconds, with no human intervention.

Programmatic financial advising

In 2010, Jon Stein, a 30-year-old entrepreneur, launched Betterment[40], one of the first "Robo-advisors" for individual investors. With Betterment, individuals tell the system their financial goals, and the system recommends a diversified portfolio with automated investing and tax-saving strategies. Those include portfolio rebalancing, tax loss harvesting, and asset allocation—all at significantly lower costs than working with a human, financial advisor. Since its introduction, almost every major financial institution has developed or acquired its own Robo-advisor.

Programmatic media advertising

Programmatic media uses technology to buy ad inventory, while traditional media buying still relies more on human negotiations. The traditional media buying approach is more inventory-centric, whereas the programmatic approach allows granular, user-centric targeting based on data, particularly behaviors. This programmatic approach enables advertisers to track the performance of campaigns in real time and make necessary adjustments.

Programmatic media traces its origins back to October 1994 when AT&T purchased the world's first web banner ad from_www.hotwired.com to promote its new technologies. The success of that first buy led to huge demand from advertisers. In 1996 firms like DoubleClick (now Google Ad Exchange) used ad servers, the first elements of the programmatic ecosystem. By 1998, the first digital advertising networks started to emerge. In 2007 ad exchanges, dominated by Google, built electronic auction software that automated advertising sales and purchasing—an advantage they maintain to this day.[41]

The Rise of Digital Business, IoT, and AI

People had already been talking about the prospect of an Internet of Things for decades, when in 2012, Gartner's first Hype Cycle for the Internet of Things defined IoT as "the network of physical objects that contain embedded technology to communicate and sense or interact with their internal state or the external environment."[42]

All the buzz about IoT triggered an important new line of research for us in 2014: "Digital Business." Besides advising companies about opportunities in e-commerce, Gartner saw that billions of IoT devices and the data flowing from them, would eventually reshape most companies. From our work on Digital Business and IoT something new arose: the idea that machines could become customers.

Concurrent with the development of IoT, we saw another major area of rapid technological advancement – Artificial Intelligence (AI). Uh oh – did we just use *that* term? The one that defies a strong definition because it's such a moving target. Computer scientists have been arguing about what is and isn't AI ever since the Dartmouth College conference first used the term in 1956 [43]. These days what most people are indirectly referring to is a collection of novel algorithms that enable advances in machine learning and pattern recognition. The particularly exciting progress of recent times has come from modern variations of simulated neural networks (NNs), particularly "deep learning" multi-layer NNs and generative adversarial NNs. The advent of chatbots based on GPT large language models has improved the way these systems can interact with humans, to the extent that a wide range of new applications now seem in reach - such as fully automated customer service and support. Rapid progress has been possible because we have so much computing power available today and so many digitally captured and stored example patterns with which to train the NNs. If you want a machine to do something like recognize and count cats in videos, you train a neural network to do it by feeding it many, many examples – and today we have the cloud and petabytes of cat videos. There are widely available developer APIs, tuning tools, testing tools, and specialized forms of microprocessors that can do the math of NN computation far faster than conventional CPUs.

The next milestone was at the 2015 Gartner IT Symposium conference in Orlando, Florida when Don introduced the idea of Machine Customers to our clients in a 30-minute TED-style presentation to a packed audience of 800 people. His presentation led us to start thinking hard about the idea of Machine Customers and their implications for our personal lives, businesses, and society. That talk led to a steady stream of research notes and presentations from the content group on the topic and the growing conviction that we had a big idea on our hands – maybe even a book.

A Future with Machine Customers: How Can They Help Us?

Humans and machines have a long-shared history. As machines have become more sophisticated, they have freed us from mundane tasks like weaving fabric, producing food, and transporting us anywhere (even outer space). It is inevitable that machines, now endowed with increasing levels of intelligence, will continue to do even more for us, including our work as customers. But do we really want to start delegating our shopping to a custobot? After some reflection, we say, yes. After even deeper reflection, we say, hell yes!

Of course, you may enjoy shopping some of the time. Maybe you enjoy looking for clothes, or going to the bookstore, or buying a special gift for a friend or partner. But you probably don't enjoy shopping for toilet tissue, tires, or life insurance. And you almost certainly don't like making business purchases - a recent Gartner survey found that 72% of B2B purchasers prefer a representative-free buying experience.[44]

The fact is shopping is WORK. That's why, sooner or later, you are going to start delegating some of that work to a machine. For the most part, shopping is not fun, not a good use of your time, and because you are human, not something you are especially good at doing. People are lazy. We forget to buy things. We don't negotiate very well. We often get played by brands that sell us things that make us feel worse.

Still not convinced? Let's look at another scenario, of how custobots might change B2B sales. In Chapter 1, we showed you how the personal life of Jessica, the time-starved journalist, will benefit from a custobot. Now, let's consider how Machine Customers will improve the working life of Jessica's sister Emily, the general manager of a large manufacturing company.

At Home: Every Consumer Would Have a Butler If They Could Afford One — Right?

Leaving Home: The Autonomous Vehicle

"Waymo" - let's go to work," says Emily, as her car door shuts, and the dashboard lights up instantly. The vehicle is fully charged, having made a late-night stop at a charging station while Emily slept to take advantage of low charging rates.

It's 2030. Led by Tesla, almost every automotive manufacturer developed a self-driving system in the 2020s. Emily's car is a supercomputer on wheels, with the ability to make transactions on her behalf from obtaining insurance to ordering new tires- and paying for them with its own credit. We're not alone in thinking of the car not just as self-driving, but as a leading a nearly autonomous existence. As early as 2016, Elon Musk outlined Tesla's plan to give cars the ability to hire themselves out for people other than their owners to use on an as-needed basis, a kind of mobility Airbnb[45]. Now, Tesla and most major car companies are creating vehicles that can communicate with the manufacturer or even the dealer about when it's time for repairs. But why stop there? Why couldn't a car top itself up or drive itself to pick something up from the convenience store? There are lots of possibilities surrounding the vehicle itself, and every car manufacturer or business adjacent to auto manufacturing with whom we've shared this concept has told us they are thinking about this today, and they think this future is very real.

Commute To Work: The Intelligent Road

As Emily sips her coffee and tries to get through the emails that piled up overnight (custobots won't solve every problem), her car speeds silently to work, guided by intelligent systems embedded in the roads and the surrounding infrastructure.

The road is perfectly smooth, very different from the roads Emily remembers as a kid. That's because roads now request their own repairs – and repair crews respond because each section of road pays for its own patching, raised from its hard-earned tolls. Companies such as Cisco, Acuity, and GE have been making systems for smart cities for years. These kinds of devices monitor what's happening on the roads, and they could become part of a system where the road itself could request its own repairs. The point here is that anything that can be instrumented and connected to the Internet, has the potential to become a customer - even a road. It's important to be broad-minded.

Stop By the Factory: The Industrial Robot

After reading an alarming message from one of the factory bots, Emily decides not to go straight to the office. "Hey Waymo, Let's go to the factory," she says. The operations team recently installed a new set of robotic assemblers, and she wants to check on the progress of their start-up procedures.

Inside the factory, most things are operating well. Factory robots schedule their own repairs these days. This has been in the works for a while – after all, even back in the 2020s IoT connected jet engines could alert airline operators when something was not right. Sensors in physical objects transmit information to a system that can predict what might happen next. Sometimes a robot needs cleaning after a hard shift, so it might order its own cleaning or order replacement parts. And, from time to time, it orders upgrades for itself, to improve its software or make it more secure. Emily has heard that they will soon start negotiating for things like insurance, as there are consequences when a machine malfunctions on a production line – both safety-related and economic.

We've spoken to insurers who are interested in the idea of insuring robots. Wolfgang Hauner, Head of Group Data and Analytics at Allianz Life, said the company is already exploring these scenarios, starting with insuring autonomous cars, trucks, and other vehicles. Allianz believes that autonomous vehicles will operate with high risk for the time being as the technology gets better - a perfect customer for insurance. Vehicles are only the beginning.

> "*It can also apply to other physical things, not necessarily cars - it can be in factories where robots will talk more and more to each other. They will produce some risks to the production line and maybe to the whole supply chain if you again think autonomous logistics*" Hauner said.[46]

Allianz and the other insurers we have talked to all admit that these are very strange scenarios. We agree. We see a very exciting future ahead of industrial machines bristling with sensors that will detect demand for all kinds of things, often before their human operators do.

At Work: Can Someone Just Make Things a Little Bit Easier Please?

Check-in With Purchasing – The Buying Algorithm

Before heading to a meeting at HQ next door, Emily makes a brief stop at the Purchasing department with a quick question. She knew that the Purchasing group had been working with an advanced system to better manage the purchasing process. This has been especially important amid all the supply chain disruptions the factory has been experiencing.

Let's get a little more abstract by shifting from physical robots to buying algorithms. We said earlier that Machine Customers are not always physical objects connected to a platform. Virtual Machine Customers may be working for the purchasing department. These buying algorithms already exist today in supply chain management systems, but often they perform very basic functions like replenishment. However, as our workplaces are advancing, with more and more sensors that detect more and more information about what is going on, algorithms will become more sophisticated as the sensor inputs to them improve. So, the buying algorithm could buy physical supplies, but it also might buy labor, perhaps hire gig workers, contract Business Services, or buy energy.

Two purchasing technologies are already emerging: Autonomous Procurement and Autonomous Sourcing. Autonomous Procurement uses artificial intelligence (AI) bots with robotic process automation (RPA)-like capabilities. These bots work with large pools of data to fully automate the procurement process for one or more categories of spending. Autonomous Sourcing is a next-generation e-sourcing application that adds AI capabilities to reduce the cycle time to execute sourcing. It automates the execution, negotiation and award decisions of routine spend, allowing the sourcing staff to focus resources on more strategic negotiations and category management activities.

For example, Datapred, a digital twin for energy and raw materials suppliers, uses machine learning to recommend optimized purchasing strategies and generate related financial risk reports based on commodity or raw material price predictions and organization-specific internal constraints. After analyzing historical pricing, market data, internal operations data and other available contextual data, as well as internal constraints (such as available warehouse storage, capital, and production plans), the system makes "buy" or "defer" recommendations.[47] Evisort, a contract management platform, has added a large language model (LLM) to its software that negotiates contracts automatically and redlines proposed changes.[48]

Companies such as Pactum are emerging as early leaders in developing AI negotiation tools and Agentic AI, experiencing a significant increase in demand and acceptance over recent years. These tools are starting to prove that people are comfortable chatting with a bot about business terms, even going as far as allowing AI agents to handle sophisticated tasks. Pactum CEO and co-founder Kaspar Korjus said, "Pactum's agents are moving towards broader objectives like running efficient procurement or saving money, capable of engaging in strategic conversations and demonstrating very advanced thinking, let's say on human levels." A key differentiator for these AI agents is the increasing level of authority they possess; as Korjus emphasizes, companies are building "an AI who has authority power." For both client procurement teams and suppliers, the most compelling evidence of success comes from seeing concrete savings on their own deals and experiencing the process benefits, such as less emotional interactions and the creation of win-win opportunities, particularly benefiting smaller suppliers who previously received less attention from their human customers.[49]

To be clear, Machine Customers will not be specifying and contracting for large-scale, unique, or highly complex B2B investments such as new container ships or office blocks. However, they will become more sophisticated at selecting and negotiating for moderately complex goods and services, from print services and insurance to aggregates and electrical components.

Go To a Meeting: Conference Room

After her visits to the factory, Emily's car whisks her to HQ for her weekly staff meeting.

Her conference room might look familiar to you, but it's actually very different. It's 2030, and conference rooms are now smart spaces – an environment in which humans and technology-enabled systems interact in increasingly open, connected, coordinated, and intelligent ecosystems. Multiple elements - including people, processes, services, and things - come together in what Gartner calls a smart space[50] to create a more immersive, interactive, and automated experience for a target set of people or industry scenarios.

Nobody has arrived yet, but the coffee and rolls have arrived, and Emily picks up her custom coffee, which the barista bot has prepared just the way she likes it - an oat-milk latte with two sugars - and a banana muffin. It's her favorite breakfast - and as everybody on her team files in, they find their favorite breakfasts waiting too. Over time, the room has learned their preferences, including two extra napkins for Sid, who tends to be a bit messy.

Sound like science fiction? It is – but only because we haven't done it yet. Cisco, Microsoft, and Herman Miller offer a variety of office technologies that could make this a reality.

Take a Break: Drinks Cooler

Unfortunately, Emily's favorite soft drink was not part of the catering for her staff meeting, and she still has a craving for it. At a break in between meetings, she walks to a convenience store next to the factory which is sure to have her favorite iced tea in stock - the refrigerator is programmed to never run out.

We've talked to several beverage manufacturers who are looking at the idea of making these coolers better buyers of products on behalf of the store owner. Peter Schwartz, Senior Vice President – Strategic Planning and Chief Futures Officer at Salesforce, shared a similar example:

> *"So, let's take an example of Sysco, the food services company. It's not hard to imagine today a supermarket refrigerator in a restaurant is communicating with their Sysco driver, saying, 'I've just been checking with the weather service. It's going to be really hot here in the next few days. Will you deliver three extra cases of ice cream, please? And that is happening autonomously in the background. Now, at some point in that flow, it's going to pop up and check with the driver and with the owner and says, hey, I've just recommended*

to Sysco they deliver three extra gallons of ice cream. Are you okay with that? And literally, we're planning to do exactly that - being able to enable that conversation to take place."[51]

Many of the people that run small operations deal with a lot of things: staffing, stocking inventory, manning the register, security, etc. Perhaps the restaurant's refrigerator can place a restocking order on behalf of the owner. That could be of value to the owner and even to the manufacturer or distributor themselves. We see this today already in smart vending machines, like the ones offered by AutoCrib's RoboCrib[52] that place their own reorder for consumable supplies like gloves, light tools, and small machine parts.

Back Home: Can Someone Help Me Take Care of Me?

Home Fitness: Workout Mirror

After a long day, Emily makes it home safely in her autonomous vehicle. To burn off some of the stress of the day, she's going to squeeze in a quick home workout before dinner. She has one of the digital exercise mirrors that coach you through your workout to make it as productive as possible, and look after her in other ways, such as ordering new exercise clothes when hers are starting to look a little worn out, or suggesting she is looking a little pale and may want to take an iron supplement.

Companies such aslike Lululemon, Magic.fit, and Tonal already offer devices and programming to guide your home workout.[53] An exercise mirror can become a Machine Customer. You can see yourself, but it also displays a video that allows a virtual trainer (pre-recorded or live stream) to see you and coach you through your workout.[54]

Consider this: advanced video analytics powered by AI technology watches you as you exercise. It is evaluating and learning your movements to help you improve your technique, so you can reach your fitness goals. It also sees what type of fitness apparel you are wearing and how those change over time. Maybe you are losing fat or building muscle (or both), and you may need new apparel. This gives the custobot all kinds of possibilities to curate your fitness life. Health and nutrition can also be part of this relationship. Mobile applications like MyFitness Pal, Yazio, and Cronometer help you track your food intake. Could that information be linked to your fitness mirror? Very likely. The Machine Customer can then help curate your health more broadly, recommending and buying a variety of items you may need in your pursuit of health. It becomes a true health partner.

Yes, this part sounds creepy. A custobot is learning about you, so it can serve you better. But that is what a human personal trainer does and like a personal trainer, Machine Customers must see us to help us. We will be presented with many ways they can enter our lives, and we must trust them if they are going to become good buyers on our behalf.

Cook Dinner: Appliances

After Emily has finished her exercise and shower, it is time to eat the meal she has earned. Luckily, her kitchen is way ahead of her. It knew she was home and proceeded to heat the meal she had prepared over the weekend.

We are all reminded of the many tales and times gone by of the idea of kitchen appliances restocking food and supplies for us. You may have seen announcements like this:

> In 2016, Samsung announced that its SmartHub refrigerator would be able to order with an Internet of Things function that monitors the contents and can place an online order to refill shelves. [55]
>
> In 2019, Tokyo-based Shiftall announced DrinkShift, a beer refrigerator that can restock itself.[56]

As of the writing of this book, announcements like these often contain more hype than reality. The 'trough of disillusionment' in the Gartner Hype Cycle is an inevitable development stage before novel ideas like this win market success. Ambitions today are a bit more modest. We see limited automation and reordering of supplies as being the first step to more advanced restocking capabilities.

> Nespresso's Essenza Plus pod coffee machine can send an alert to your mobile phone suggesting it is time re-order capsules or when it is time to descale the machine.[57]
>
> LG offers a line of connected appliances, including a fridge, dishwasher, and washing machine[58], integrated with Amazon's Dash Replenishment Service so that you don't run out of laundry detergent.[59]

But what about cleaning the appliance. In the case of Whirlpool, perhaps the app is integrated with TaskRabbit, a gig worker site for a variety of jobs. The microwave oven orders a cleaning, communicates with the home's smart lock to enable a gig worker entry to the home, giving the person one hour to steam clean your oven. Of course, this could work just as well for any type of connected appliance or system, like a heating and air conditioning unit.

Clean-up: Household Robots

Now that dinner is done, Emily notices that the kitchen floor could use a bit of cleaning. She asks her household robot to sweep and clean the floor while she gets ready to relax.

House cleaning is a chore we are all more than ready to give up to machines like iRobot Roomba or similar that sell in the millions. As these devices clean your home, we believe in the future they will start ordering consumables such as cleaning fluid. For example, iRobots' Braava Robot Mop cleans your floors automatically[60] But that's not all. We see the possibility that when the machine is idle, it can be hired out to your neighbor who also wants clean floors. Why not? In today's sharing economy, if something is not being used, like your lawnmower, someone else can rent it through FriendWithA.[61] There is nothing to say that that couldn't also happen for household robots. Let's take that thought further. What if these robots could *hire themselves* out to other people, make their own money and spend it?

Relax: Smart Speakers with Intelligent Assistants

As Emily settles into her comfortable chair in front of the TV, she asks her voice assistant to play the latest episode of her favorite TV show, but Alexa reminds her she didn't like the last two episodes and recommends a new program that has gotten much better reviews.

During the global pandemic, consumers made extensive use of voice assistants to interact with technology during lockdowns and get them to do things on their behalf like reminders, alarms, playing music and movies, placing orders, checking on the status of a shipment, etc. The promise of these voice assistants, like Siri, and Alexa, is they are constantly learning. We believe that as they learn more and more about you, they will do more and more for you.

Conclusion

From their humble beginnings in 1950's England through to the rise of the internet and e-commerce, Machine Customers have been evolving in the shadows for a long time. As we add the increasing capabilities of the Internet of Things, Artificial Intelligence, and Digital Business their role will become more visible. The stage is now set for the rest of the book – exploring the business, human, and societal implications of Machine Customers and why we believe they represent the final frontier of growth on planet Earth and beyond.

Key Takeaways and Actions for Executive Leaders

Customers don't have to be human.

Action: Add Machine Customers as an emerging customer segment in your long-range strategy planning.

Machine Customers are now capable of acting on our behalf.

Action: Think about ways machines could act on behalf of your customers, even if it is simple automation of one or two tasks in a customer journey.

Machine Customers will free us from the mundane, just as technology has done for thousands of years.

Action: Identify the drudgery in your customer's lives. How could a Machine Customer take some of it away?

Chapter 3

The Custobots ARE Coming

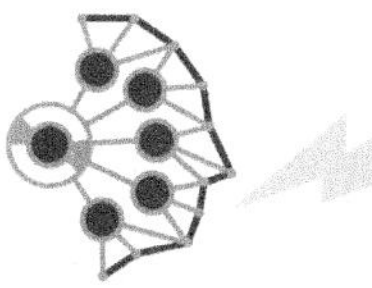

The Custobot is always right.

The "why?" of a new business technology can be fuzzy at first - but a closer look at the underlying forces will make it clear

Machine Customers Will Often Be Better Than Human Customers

If you take a cool, dispassionate look at your business, you're likely to find that you should fire your worst customers[62]. The truth is that some customers just aren't very good at their job. They dither, they waste a lot of time, and they send stuff back. In B2C situations, their cost-to-serve may exceed their profit potential. In B2B situations, endless rounds of negotiation and contract clause discussions can inflate your working capital and tie your organization in knots. Smart businesses rank and yank their relationships so they can focus their energy on their most profitable or promising customers.

Over the next few years, as technology improves, smart machines will become more capable and effective at being customers than their human counterparts. Soon, smart companies will seek out more Machine Customers and may start dissuading or even firing some of their human customers.

Efficient, Logical, Scalable - All the Things You Aren't

How Machine Customers Will Unlock Hidden Growth

Human customers are often suboptimal. We miss appointments. We forget shopping lists. We don't know how to search and find what's good for us or best for us. Human time and attention are finite and fragmented. What we hope to do or intend to do gets derailed by something else. "I wish I remembered to order those flower bulbs for the garden – oh well, the planting season has passed – maybe next year."

Figure 2 - A Depiction of The Purchasing Behavior Difference of Machine Customers

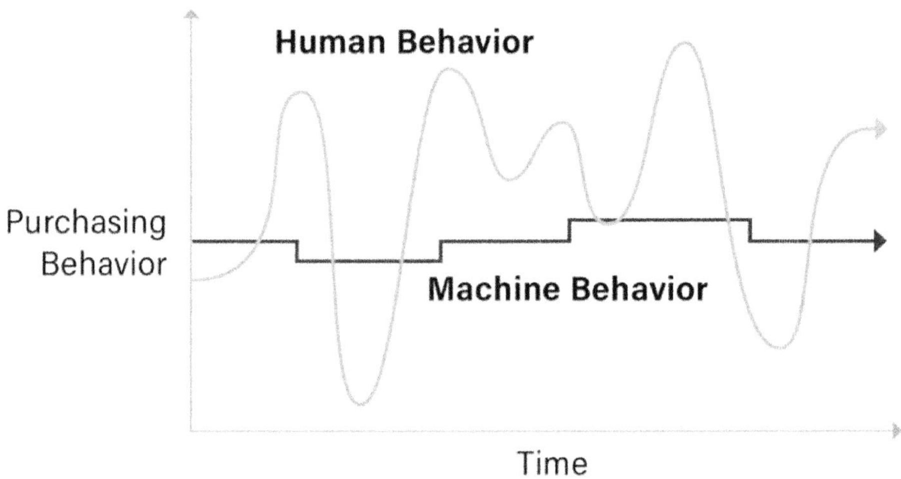

As illustrated in Figure 2., machines have several characteristics that will make them better consumers than people:

Machines are observant, tireless, and exhaustive researchers.

Machines don't forget.

Machines apply logic and reason over subjectivity and emotion.

Machines are also often more cost-effective than people: they can run 24 hours a day and they are replicable. If we need more, we can make more, using mass production methods honed to perfection over the century since Henry Ford built his River Rouge plant. Soon, we won't be stuck with randomly variable customers – we'll be able to make consistent customers who will cost less to acquire and do a better job for us.

For example: Mark has a letter from his dentist reminding him that he is overdue for a check-up. It has been on his desk for a month. He is not a reliable customer of the dentist. *Quelle surprise.* What if he could just say, "Alexa - could you manage my dental for me?" In reply she might say "sure, Mark, who's your dentist?" and then, "OK - I found them - I can take that over for you." Sadly, that service does not exist yet. However, if you have any high-level appreciation of the tech world, you probably think it is not *that hard* to do and it cannot be very far away.

But hang on a minute – we just said that at least part of being a customer is "work". That can be a challenging notion. We like to focus on the idea of shopping as recreation. Historically, businesses had to seduce us into doing this work - marketers are forever promising that shopping will be easy, fun or fulfilling. Our over-rehearsed mind's eye turns quickly to a scene of spending our hard-earned money in a swanky shopping mall as a pleasurable activity. That may be true for some categories of purchase at some times, but it is not true for all purchases all the time. In fact, shopping-as-leisure is the exception – and not just when it comes to dental appointments. For example, Thilo Koslowski, former CEO at Porsche Digital, a wholly owned subsidiary of Porsche AG asks these questions about today's cars:

> *"Why do you have to slide your credit card into the gas pump? And take it out? Why doesn't the car communicate automatically with the pump and pay for the gas? And determine how much gas is exactly needed because the destination was entered in the nav system? The car should even seek the cheapest gas station based on its understanding of how much gas is needed to make it to the destination. Why is this not happening?"*

Like buying gas for the car, most shopping isn't fun at all. Think of all the things you buy that are not fun purchases: utility bills, insurance, bleach, potatoes. We might enjoy shopping for single malt whiskey, TVs, handbags, and fine fragrances; we do not enjoy shopping for car tires, toilet tissue, printer paper, and lightbulbs.

To be sure – this is subjective. Some people enjoy shopping for basic groceries when it gets them out of the house and away from three small screaming children. What about complex things like buying a home? A local property portal in Singapore called Mogul.sg developed an AI agent called Maia that can review more than 100,000 home listings on major housing sites so it can make educated recommendations for potential homebuyers. The AI agent can also discuss ideal viewing dates via WhatsApp, send a calendar invite, and if the viewing is successful, remind the seller's agent about the binding agreement and fees once the sale is completed.[63]

But for many people and businesses, being a customer is WORK, and a lot of it is pure drudgery. Just think about all the jobs to be done, like mowing the lawn, as illustrated in Figure 3.

Figure 3 - The Wider 'Jobs to Be Done' Coverage of Machine Customers

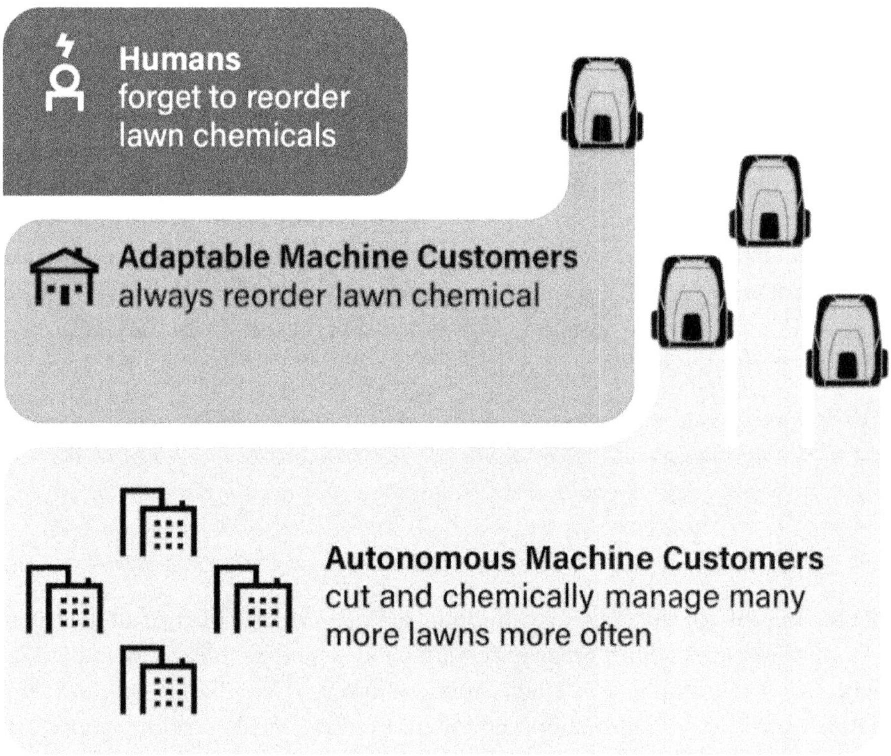

In recent decades, companies have often used technology not to automate customer service, but to offload it onto their customers. Self-service supermarket checkouts require you to scan your groceries, while self-booking travel sites make you hunt for your own flights, saving them money and saving you... what, exactly? Generally, either from doing something you are better at, or something you enjoy more or both. Still not convinced? Here is how we can be sure – rich people don't do it. They delegate shopping to their housekeeper, butler, or personal assistant. It is worth remembering that throughout business history, fortunes have been built by taking something that only a few rich people can have and finding a way to make it a mass market – from tea to quick service restaurants to low-cost aviation. We believe shopping is next.

The Machine Customers Are Already Being Dispatched to Your Location

How will your business acquire Machine Customers? To begin with, you may want to try unlocking your front door.

Have you ever encountered a website that required a 'Captcha' before letting you in deeper? Most of us have. We take a little test that asks us to click away on picture squares or decipher squiggly letters and declare, "I am not a robot". The Captcha technology website says:

> *A CAPTCHA is a program that protects websites against bots by generating and grading tests that humans can pass but current computer programs cannot.* [64]

There are lots of good reasons why a commercial website might want to protect itself against bots. For example, to protect the integrity of online polls, stop brute force attempts to crack passwords, or prevent competitors from copying today's prices. But here's a thought – what if some of the bots are customers you should *want?*

Screen scrapers predate the commercial internet, but when companies started showing themselves worldwide on the web in the 1990s, their use exploded. Today, one of the early things many new programmers are taught is how to extract data they want from a web page intended for human eyes – such as sports scores or stock prices. Libraries of well-refined code that can do this clever task well are freely available on GitHub and other open repositories. Using these tools, even an inexperienced coder can easily put together an app that interacts with a web page and "pretends" to be a human user.

Now think about this. Many people who have customer work they regularly need to do with businesses online can do some coding these days. But if your site administrators have decided to use Captcha, the code those customers write for themselves could be locked out. The chances are good that your website is *already* receiving visits from Machine Customers. In our research, we learned of a few forward-thinking organizations that are building their own bot frameworks to negotiate with incoming bot requests.

Of course, there are good reasons to block even well-intentioned screen-scrapers. It can be a technologically inefficient means of interaction. It can be insecure, and it can cause site reliability problems. It might be better if your company provided APIs that allowed programmers to access the functions they need in a formalized, registered, metered, orderly, secure, and professionally managed way. In fact, we may look back on screen-scrapers as the protozoa of the Machine Customer age - a crude animal from which more sophisticated fauna will evolve.

Indeed, we can already see that evolution happening in that rainforest of Machine Customer biodiversity called eBay. The vast virtual marketplace has so many small traders and amateurs buying and selling things that Machine Customer bots are already mutating and evolving. One long-time King of the Jungle is sniping software - a bot programmed to swoop in at the very last second of an auction. In 1999, esnipe.com first freed users from having to spend time watching the countdown clock, and today there are many competing solutions that do the same thing. Other companies might try to shut down the snipers, but not eBay. Indeed, it supports their evolution with APIs that allow programmers to create professional and high-quality Machine Customers.

Ebay's "Buy API" capabilities[65] help shopping bots perform research to make good buying decisions. For example, they enable programs to:

- Compare listing details, such as price, shipping, return policy, and others.
- Search for products using standard identifiers like UPC.
- Show product rating, reviews and seller feedback scores.
- View item images, sizes, colors, and other item specifics or variations.
- Retrieve relevant item and product recommendations to cross-sell and up-sell additional items on eBay for buyers.

The functions those APIs offer would be a good starting list for any company to consider when it decides to engage with the world of Machine Customers.

However, the early manifestations of online shopping bots have not always been seen as a positive development. Sometimes regulators have been called in to thwart them. This is most obvious in the case of sports and show ticket sales. Ticket bots have become such powerful tools in the hands of reselling intermediaries that laws were passed to curtail their use. For example, the US Better Online Ticket Sales (BOTS) act was signed into law by President Obama in 2016.[66] The problem was not that the bots didn't work, but that they worked too well, when set loose by scalpers. But even now, this technology is still available to individuals.

Next time you are trying get a ticket for a hot show or a big game, on the day they are released, try Googling for "Ticketmaster bot software". It may not be illegal, but *should* you use it? Is it fair? You will have to make your own decision, but the tech genie it is out of the bottle and its effects are extending into other industries. The athleisure shoe segment has been challenged by the arrival of 'sneakerbots' that attempt to swoop in and buy up limited edition inventory online. Nike has had to take steps to try and control sneakerbot-enabled buying[67] As the use of shopping bots expands, new solutions will be required to manage the way vendors release inventory for sale. For example, Queue.it offers software to *"deliver a fair online experience without crashes or bots"*.[68]

Digital Channels Will Evolve to Become Machine Customers

Have you ever thought about the shopping basket icon on a website? When ecommerce was invented in the middle 1990s, it was so conceptually revolutionary it needed to be explained by reference to physical retail concepts that users would immediately understand. That's often the way with technology revolutions. The icon for email is a paper envelope and the icon for saving a file is a floppy disk. We always frame tomorrow's technological possibility in today's familiarity and experience. So it will be with Machine Customers.

We have no doubt that the emergence of Machine Customers will be obscured in the blind spot from which they emerge – digital channels. For the past quarter-century, the pattern of tech-enabled market evolution has been electronic channel proliferation. Sales and marketing professionals discuss customer journeys that crisscross integrated multichannel or 'omnichannel' surfaces comprising web, mobile, email, IVR, kiosk, and others. Perhaps the most confusing of the transitions will be from voice-enabled or conversational commerce to the Machine Customer.

This transition is already happening with Amazon Alexa. First, you order online, and Alexa's notification light alerts you about deliveries. Then you start asking Alexa vocally to reorder stuff. Soon enough, Alexa pipes up with messages asking if you would like to set up a subscription, or she proposes items from your wish list that are on special offer. Recent advances in generative AI are likely to bring this all to life more quickly than we initially anticipated. For example, already, a ChatGPT feature has been created for Alexa called Alexa+.[69] ChatGPT can provide powerful shopping support such as suggesting lists of the best brands and models of an item, or providing a list of steps to follow in a particular negotiating situation.[70] We are seeing more and more digital commerce leaders like Walmart, Expedia (Romie Assistant), and Zalando announce or embed generative AI capabilities. [71] Walmart for example is anticipating the need for industry protocols that allow a variety of AI agents to transact with retailers.[72] This will significantly boost the quality of co-customer advice and support that virtual assistant Machine Customers can provide.

Before you know it, a rather smart virtual assistant will be in your life, chatting to you about what you should buy next. Already, Alexa and other chatbots are starting to be involved in product selection and prompting purchases. Maybe they aren't quite custobots yet, but it is as if the Machine Customer is emerging out of the digital channel soup like the liquid metal T-1000 Terminator character in the movie "Terminator 2: Judgment Day." So, are they a channel or a co-customer? It's too early to tell, but we are sure to find out soon.

As channel tech becomes smarter, it does more and more work helping and guiding the human. The evolution starts with little things like pre-filling forms. It moves on to curating and reducing choices to match personal tastes. From there, it goes on to make proactive suggestions. At some point, the tech is no longer a pipe. It is an intelligent co-customer. More and more of the decision-making is happening in the machine – less in the human. Eventually, we reach a tipping point.

It's not hard to envision that a dialogue that starts with "Alexa reorder dog food" eventually reaches a moment when the intelligent assistant asks, "Would you like me to take care of all the pet supplies your dog needs?" If you say yes, the next question is "what's the monthly budget I can use for that?". At that point, the machine has been designated as a customer on your behalf - just as if you had a human housekeeper. The machine can search for the best pet foods. Perhaps it will adjust amounts according to dog walking activity data from a Fitbit-like IoT collar and smart pet feeder (PetSafe's 'SmartFeed' already exists[73]) and listen to your dog's night-time breathing. When other pet owners discover new chew toys, the machine would order the occasional surprise – all within the budget.

Machine Customers will have a counterpart: The Machine Seller

As Machine Customers develop in their capabilities, so too will their vendor counterparts – a Machine Seller, or what we call a "sellerbot." If custobots are hired by human customers to represent their interest when interacting with third parties, think of sellerbots as machines that are built and managed by organizations to sell to and serve human customers (and eventually other custobots). In discussions with Gartner colleagues and clients, we had to account for organizations that want to use machines to sell to others – including Machine Customers. We'll discuss sellerbots more in Chapter 6.

The Forces Driving Machine Customers Are Huge and Inevitable

Growth is a need but not a given. Most businesses rely on the idea of growth as the central tenet of their existence. We seek growth; we set goals for growth. When growth isn't high enough, questions are asked. When growth reverses, it's a big problem. But it never just happens. We can't rely on it. We must cause it.

Economies And Societies Need Growth – It Is Who We Are

Throughout history, the scale of the economy has been somewhat proportional to the number of humans. It isn't a perfect and simple numerical relationship, of course. When birth rates go up and the population increases, it does not assure that GDP growth will immediately follow. However, it tends to create pressure for that to happen. Otherwise, resources must be spread more thinly. We can say that declining, and shrinking populations cause economic stress. Structurally that is true in countries like Japan and Italy today.

Why are we so sure Machine Customers are coming?

We need the growth.

Human population growth is forecast to come to an end later this century.[74] When that happens, where will our economic growth come from? We believe machines will become the additional customers we cannot be. They will help us improve our quality of life in ways that do not lead to the ecological damage of an expanding population. Their growth will make it possible to raise standards of living in a sustainable way, helping the elderly and cleaning up our legacy of environmental messes. By helping us find and buy what we really need when we need it, they will help us avoid excess, damaging overconsumption, and waste. But how do we know Machine Customers won't just cause us to consume more stuff? We don't. Initially, they may have both good and bad environmental effects (which we expand on a little in Chapter 11). However, we are optimistic that in the long run, they will help us live more sustainably. For one thing, society will want it that way. As IKEA's chief sustainability officer has said, we may have already reached "peak stuff".[75] We are saturated in fat and sugar to lifespan-shortening excess, head spun by endless attention-grabbing dopamine hits, driven round and round in too many cars, while wearing low-grade fast fashion that goes almost straight to landfill.

We think that in the end, most people want a *better* life, not a "more, more, more" chore bore life. And even if Machine Customers were programmed to keep convincing humans to consume more, we would soon reach planetary limits. Many natural scientists are reaching the conclusion that the earth has physical limits, and we are closer to reaching them, requiring a change of approach. If we humans are too irrational to make that shift alone, perhaps more logical Machine Customers could help us.

Machine Customers Are Part of The Next Giant Leap for Capitalism

Capitalism is adaptive. As Paul Mason observes in his book Postcapitalism[76], it always seeks and finds the next realm to monetize. It moved from land and food to credit cards and education. It surfs from technological wave to technological wave as Nikolai Kondratiev[77] first foresaw in the early 20th century.

The leap to Machine Customers will be a big part of the 21st-century digital technology golden age of economic progress and wealth creation that Carlotta Perez laid out in her landmark book Technological Innovations & Financial Capital.[78] The financialization and control of the Machine Customer economy will help sustain growth, enabling it to break through the barrier of human population stasis. We believe that by 2028 machine customers will be recognized and measured by economists as a factor that contributes to the growth-reliant model of a stable society. It might help us all live better lives, and it stands a chance of helping us control the dangerous aspects of excessive consumption that are poisoning our planet and may have started to reduce our lifespans.[79]

The Digital Giants Will Simply 'Make It So'

Amazon and the other big technology corporations think a long way ahead. They create the tools and lay the groundwork for the next big tech wave and the one after that. They know that Machine Customers will be a major battleground, and they have already staked the territory.

The Technologies Needed for A Machine Customer World Already Exist

As we wrote in a previous book Digital to the Core[80], in the end, "every industry will be digitally remastered." That means its products and services will be substantially reinvented by the direct inclusion of digital technology. Autonomous cars, robot lawnmowers and vape sticks are already tearing up the competitive strategy rule books of industries. Did you know that Apple Watch alone already outsells the entire Swiss watch industry?[81] If you are a Silicon Valley VC, that factoid is just 'so yawnsville'. It's yesterday's disruption. The question is where these IoT things will lead us. What's the *next* S curve to jump to?

Technology enables future business growth in waves. A technology yields new rapid growth and wealth creation itself for a while, and then eventually, that slows as markets saturate. But another wave follows, built upon the infrastructure of the last. PCs and modems begat the web. The web could only come to pass once everyone had a computer on their desk and an internet connection. The web begat social. We could not have Facebook, LinkedIn, and the rest until everyone was comfortable with "web surfing" (quaint, that old term – isn't it?). IoT technology grafted into many products will create a massive new infrastructure and what we will build next, on top of it, will be a Machine Customer economy.

The evolution of Machine Customers will also depend on AI-based pattern recognition. When someone holds their phone up in a bar using the app Shazam to listen to the music and recognize the song being played – that's pattern recognition of an audio waveform. When you speak to Alexa, Google Assistant, Siri, Tmall Genie, or Bixby, the waveform of your speech is recognized. Once the individual words are found and identified in your speech, another AI technology called natural language processing is used to make some sort of sense of it. What's the sentence structure? What are the clauses? What's the noun phrase, the verb phrase? Was it a statement or a question? And so on.

Image recognition is another massively valuable area of AI pattern recognition advancement. Most people these days take it for granted that police, city authorities, and even your local gas station use ALPR (automatic license plate recognition technology), but of course, it goes far beyond that. Tesla cars can recognize people, bicycles, road markings, streetlights, and many other features, in real-time, by analyzing the images taken from onboard video cameras.[82]

These incredible technologies were on the whiteboards of scientists and the keyboards of sci-fi authors for decades. Now they are here. Alexa can "hear". A Tesla can "see." Moderately smart, if not actually intelligent, machines have arrived, and with these capabilities, they can take on the work of customers. They can hear what you want, use search to find versions, compare and choose and then perhaps see by observation whether a chosen product really worked out. Did you like it? Perhaps all they will need to do is watch your face when you open the package to see if you smile. It's not hard; there's an API for that.[83]

But machines are not going to pull up to the checkout at a store in your local shopping mall anytime soon and sigh when they must join the line. So "where" will Machine Customers do their shopping? Online, of course. It is the cloud and the amazing commerce platforms it has enabled that have already created the massively capable virtual market space in which the Machine Customers will operate.

Digital commerce in all its forms – web, mobile, social, video, chat, and all the rest have, in the space of just twenty years or so, created colossal marketspaces of a scale and complexity that no twentieth century business thinker could previously have imagined. Amazon, Alibaba, eBay, Craigslist, Etsy, JD.com, Shopify, Mercado Libre, Jumia, and others operate huge platforms and clearing houses that match buyers with sellers. They provide product and price data to allow comparison shopping on more features and data points than ever before. You could spend a lifetime doing nothing but shopping in these endless virtual malls and barely scratch the surface. But we believe increasingly you will not wish to visit them at all because they will become too complex for humans to understand.

Cloud-based platform technologies are key to the rise of Machine Customers, and as they become more complex, the human customer will become overwhelmed by choice and confusion. What keywords should I use? Which filters after that? How should I organize the search results. Can I trust those reviews? Why can't I ask it to compare based on battery life instead of on price?

Do any of those issues seem familiar to you? Perhaps they even feel like work? It seems *inevitable* to us that Machine Customers will reach into these deep and complex markets on our behalf. They will do that for us via APIs, and the digital giant companies know this because they are already providing those computer interfaces.

Amazon Dash Replenishment Services have existed for several years, though you may not have heard of them. They are the developer APIs behind the more well-known but unsuccessful and withdrawn 'amazon dash button' gadget. As the developer site says:

> *"Dash Replenishment Service" offers zero-click reorders for products linked to your device. The easy-to-use APIs allow you to integrate with Amazon ordering and fulfillment so that your customers never run out of products they need."*[84]

Let's examine that clause by clause. *An API to enable online replenishment of products…* means buying stuff. *Linked to your device...* that means connected products like washing machines. *With zero-click ordering...* means no human finger clicks on the mouse button. At this point, the machine is no longer just a channel-it's a customer.

Amazon, Tencent, Alibaba, and all the other digital giants are taking the lead in building the Machine Customer ecosystem, but they won't be able to do everything. Business history is full of corporations whose power and overreach as conglomerates eventually led to sloth, malaise, and collapse. We believe many other companies will copy the tech capabilities of the digital giants, marry those with their conventional strengths and become what we call "digicorps". In fact, it's already happening: Walmart now has sufficient tech capability to sell it to other retailers via Walmart.io. And Volkswagen is building a 10,000-person digital expert workforce to go head-to-head with Tesla in the battle for the electro-digital autonomous services-model future of the car industry.[85] A quarter of the employees at Goldman Sachs are in software engineering [86]. At some point in the future, we believe the major corporations left standing will become as tech-capable as the digital giants, even as those digital giants invade and take over more "traditional" competencies – such as mass manufacturing of cars. We have called that future point of market evolutionary balance *techqulibrium.*[87]

Thilo Koslowski describes the relationship between the digital giants and the auto industry this way:

> *"I believe many of the traditional automotive companies are somewhat stuck, unfortunately, trying to understand how that Machine Customer benefits their brand, their operations, but they are missing the bigger picture. The challenge is that the big technology leaders, like Amazon and Google, basically tell automotive companies to accept their offerings in one specific way without understanding the needs necessarily, of these automotive companies. They basically tell automotive companies: hey, if you want to have our CarPlay functionality in your car, this is how you have to do it, and this is how it's going to look. There's no way for you to customize or differentiate. But it is exactly this differentiation that is critical to an automaker's success."*[88]

The Machine Customer future, despite being so highly digital, does not automatically 'belong' to the digital giants and dragons. Others could have pivotal roles to play if they seize their strategic opportunities quickly enough.

Currently, less digital, more conventional corporations must understand that the digital giants will open the Machine Customer world up in a big way for one key reason: *because they can.* It is important to appreciate how core it is to their purpose and culture to explore all the strategic possibilities that digital technology advances open up. They dare not leave a stone unturned for fear that one of their direct competitors will find a big advantage there. For example, in 2023 many of the tech giants rushed into a frenzy of adding large language model generative AI to their services - a move that is greatly advancing machine customer capabilities. In 2025, that same group is rushing to build Agentic AI into their products.

- Microsoft added generative AI buyer support features to Bing[89]
- Google added generative AI into its online shopping features[90]
- Shopify created a generative AI powered custobot app called Shop.AI[91]
- Mercari launched Merchat, a ChatGPT powered AI shopping assistant[92]
- OpenTable used a ChatGPT plug-in to create a Custobot that can help plan a dinner[93].
- Expedia added a ChatGPT plugin to create a Custobot that can help plan a holiday[94].
- In Singapore, Lazada created a ChatGPT powered shopping assistant called LazzieChat[95]

The pace of competition amongst the digital giants quite often reduces fat margin opportunities to thin commodity plays within a few short years. As they find and perfect the best platform-based revenue models, they all pile into the same play.

One of the biggest plays that are becoming exhausted is the advertising revenue model. Once Google 'owned' it, then Facebook, now Amazon is muscling in.

Under such conditions, digital giants will explore any other business model possibility for monetizing technology enabled customer engagement. With Machine Customers, the prospect of a direct or indirect transaction fee offers potentially massive growth opportunity that action will be too juicy to ignore.

Control It, Before It Controls You

So big companies must move in to take their part of the Machine Customer ecosystem land grab. Companies' boards of directors must empower and fund their strategists to mark out and seize key territories. Machine Customers will be enabled by platforms of functionality and trust making. Many of the machines that become customers will be intelligently enabled for future versions of today's mass-market products. These are the future opportunity spaces that large corporations leave open and exposed at their peril. After seeing Blackberry, Kodak, Nokia, Sears, and others fall to the forces of dithering and inertia, they have few illusions that the future will stand still for them.

Most middle-sized companies won't have the market clout, risk appetite, or capital backing to create the platforms of tomorrow. Instead, they will need to get involved early so they can learn how to participate and compete in Machine Customer markets. If you make and sell, say, organic breakfast cereals, you will need to understand how machine minds will evaluate your product. This will be a complex and circuitous art in which the seller must communicate a product's value in a way the customer algorithm understands, which may not be the same as human customer influencing. Perhaps it will involve more about definitions and specifications and less about logos and images.

Start-ups will need to look for gaps to fill in the emerging Machine Customer landscapes. There will be many hundreds of these in such a complex unfolding space. How will the Machine Customer sense, connect, and move? How will it interact with humans? It's easy to see how a kitchen coffee maker becomes a smart Machine Customer, but what about a vase? Not typically an electrical product today, but could it become powered? Perhaps the vase could detect both the age of the flowers it contains and what kind of flowers it should order next based on mood, occasion, or room decor. Who will provide the smart modules that manufacturers incorporate? Is there a role for a retrofit adapter? The digital giants will bring about a custobot world, but they can't control it all. Machine

Customer evolution and innovation will be a dynamic and volatile business journey of many twists and turns. There will be ample room for small, medium, and large players to win big. But only if they are insightful, creative, and quick.

Conclusion

From using fire to cook meat to reduce chewing time to the invention of the dishwasher, we have always used technology to make our lives easier. We do it because it frees us to move on to more valuable tasks. The technology that allows us to create Machine Customers is now reaching key tipping points. The digital giants will lead their inevitable development, but they won't be able to control the entire market.

Key Takeaways

Business strategists must see this trend as a certainty.
The only real question is timing and the nature and degree of progressive impact at the industry level. In long-range planning, assume this will happen.

Action: Ask yourself what kinds of objects do customers use in the context of your products and services today? These could be very mundane if they occupy a useful demand-detecting position. Could those become Machine Customers? For example, Amazon thinks something as previously inert as a shelf could play a customer role.[96] Choose three time-horizons and assign each a probability.

Machine Customers will often be better.
Being smarter, Machine Customers will often be better at buying what's necessary.

Action: Ask yourself, what is the current level of wasted buying opportunity in your industry today, simply because humans are not very attentive or engaged in the key buying stages? How big might the opportunity be if machines acted as delegated customers on their behalf? Which Machine Customer information gathering, and decision-making stages would be the pivotal control points of these future larger markets?

The digital giants will build the first generation of custobot infrastructure.
Other very large, newly tech-transformed traditional corporations that we call digicorps will be involved. Market makers make markets – for others to participate in and profit from.

Action: Engage your competitive intelligence team to scan for examples of where the digital giants are enabling Machine Customers and serving them.

Chapter 4

Machine Customers Could Make or Break Your Business

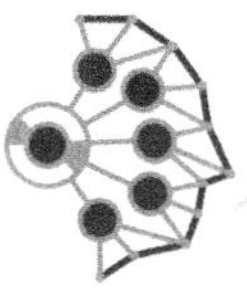

> For the first time in history a company can literally make its own customers.

Ready or not, custobots are coming to every industry– a big growth opportunity for those who seize it and a silent killer for those who don't

We Used to Find Customers - Now We Can Create Them

The founding parent of modern management Peter Drucker once wrote:

"There is only one valid definition of business purpose: to create a customer" [1]

Of course, Drucker meant a business should turn people and businesses into customers. But the arrival of Machine Customers takes his insight to a new level: What if a company really could *create* a customer?

When we look at what fundamentally limits the total addressable market size, it often comes down to one thing: the number of consumers available. The number of companies that a human population can form into B2B business customers is similarly bounded. These limits seem inescapable - but the arrival of Machine Customers might just break those limits down.

Since the industrial revolution, technology has been applied very successfully to the purpose of mass production. From the second half of the twentieth century onwards, the insertion of digital technology has made non-linear improvement rates commonplace. If we take the power of mass production, add the exponential progress of digital and apply both to the manufacturing of machines that behave as customers – then what we have is a recipe for explosive growth.

Machines that can act as customers have the potential to amplify the total size of a market. Eventually, Machine Customers may even become a market in their own right.

Machine Customer Replication Will Create Megamarkets, Mega Fast

There are already more machines capable of acting as customers than humans on the planet. Tens of billions. Yes, really - think about all the connected cars, coffee makers, smart speakers, video doorbells, factory robots, and other devices that have been on the market for several years already. These kinds of smart everyday objects already possess the processing power and capability to become Machine Customers. It will just take a while for our social norms to adjust to the idea and start exploiting it fully. If you doubt that – think about how the COVID-19 pandemic jolted us all into using videoconferencing regularly in our lives. That wasn't new technology – indeed, some people had been using Skype, WebEx, and other platforms for over a decade. What we are saying is that the mass Machine Customer opportunity is already here technologically – we just haven't taken it up yet. Or to put it another way – Machine Customers already live amongst us; we just need to empower them to act.

Moore's Law (the number of transistors on a microchip double about every two years) means that every year Machine Customers will be able to remember more usage history, listen better for more of your tonal nuances, compare more product features, and compute price discounts faster. If you think the possibility of a Machine Customer world is five to ten years away, think again. The doubling effect of Moore's Law means it will arrive much sooner than your gut tells you.

Business leaders are beginning to see the first wave of Machine Customers just ahead. About half of the CEOs we have surveyed believe Machine Customers will be significant in their industry by 2030 (see Fig.4). CEOs were surveyed by Gartner in 2019[2], 2022, and 2025. On average, the CEOs surveyed estimated that by 2030, around 15% to 20% of their revenue will come from Machine Customers. This is a remarkably consistent viewpoint, something we often don't see in our research.

Figure 4 - CEOs' Expected Rate of Arrival of Machine Customers

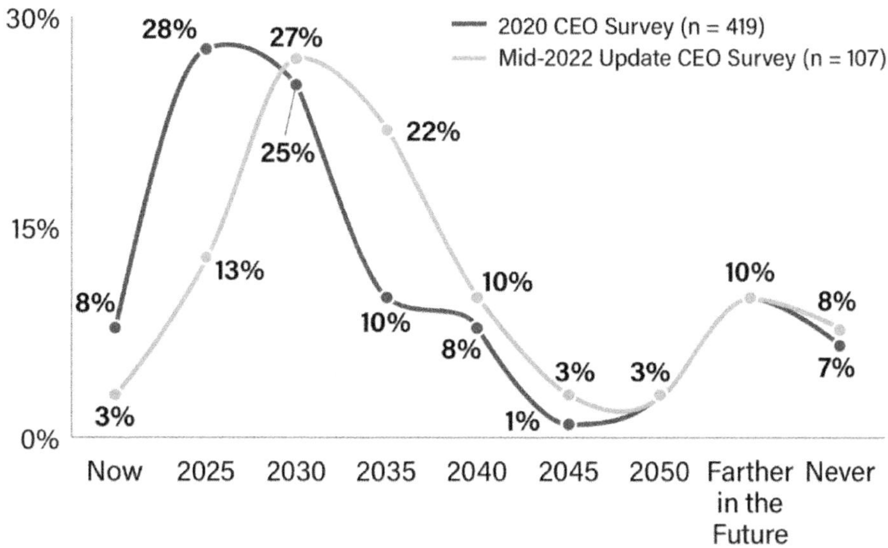

Our Gartner colleagues David Furlonger and Christophe Uzureau see Machine Customers as one part of a wider shift towards a programmable economy progressively rebuilt on blockchain-enabled foundations that are sometimes referred to as Web 3.0.[1] The programmable economy is a natively "intelligent" economic system based on a large-scale infrastructure of distributed and decentralized digital resources that supports the production and consumption of

[1] Web 3.0 (also known as Web3) enables peer-to-peer interactions with no reliance on centralized platforms and intermediaries. Users own their own data, identity, content and algorithms. They can govern the blockchain protocols they use by owning governance tokens. They participate as "shareholders" by owning the protocols' tokens or cryptocurrencies.

goods and services, enabling diverse scenarios of innovation, entrepreneurship, and exchange of value (monetary and nonmonetary) among humans and machines. David and Christophe assume:

- Machine Customers are gaining economic agency.
- Products and assets are aggregating, disaggregating and reaggregating on demand.
- Digital assets support new and more programmable types of capital, financing and value exchange.
- Fluid enterprise charters, such as distributed autonomous organizations, will replace traditional governance structures.

In David and Christophe's model, Machine Customers play a big role in generating the value that the programmable economy will yield (Fig. 5). They estimate the programmable economy as a whole (of which Machine Customers are one part) could add a staggering $163 trillion of cumulative additional value to the global economy over the decade to 2030.[i3] Even in these inflationary times, that's still a lot of money, given global GDP is roughly $80 trillion a year. How do they get there? They believe Machine Customers are going to do a lot of the heavy lifting. (Fig. 5)

Figure 5 -
Modelled Size & Growth of The Programmable Economy Over The 2020s

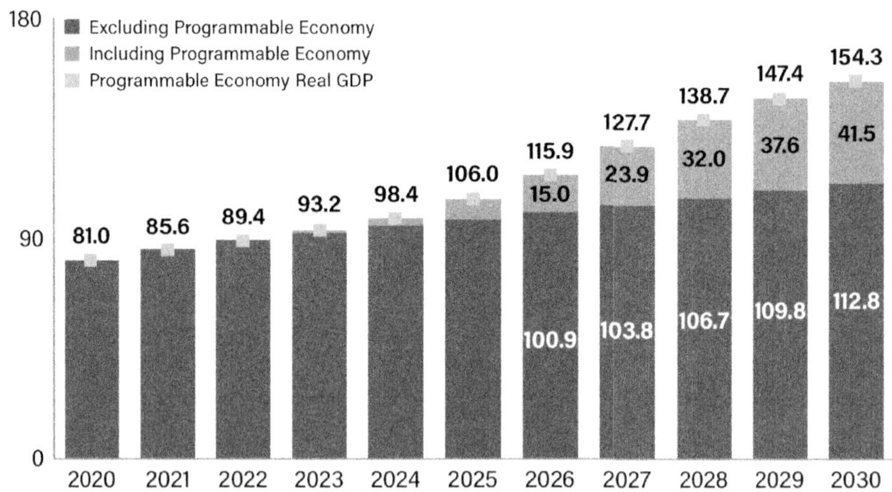

How will such a monumental transition get underway? If we are all comfortable as human customers in our current markets, how will Machine Customers displace and replace us? Even if they are better, more efficient, and save us time – we are creatures of continuing habit, and our human interactions seldom evolve at a Moore's Law pace. The reason is that in the end, human beings are good at accepting advantageous changes. After all, our predecessors gave up heating their homes with coal, clearing their streets of horse manure, and writing by hand when technology offered labor-saving alternatives. We'll sacrifice shopping.

Machine Customer Efficiencies Will Grow Markets Exponentially

Today, many markets are smaller than they could be because humans are not very good shoppers and buyers. Think about your own home right now. How many things have you been meaning to get fixed? When you eventually get around to the DIY project, will you find that your drill bit is sharp – or does it need replacing? Is the tread depth correct on all your car tires, or does one need changing? Is the repainting of the front windows of your home a few years overdue? When were the gutters last cleaned? Have you decided yet which school your child will go to next fall? Did your travel insurance lapse? Did you end up buying a crappy birthday card in a gas station on your way home – instead of the nice one your mother really deserves? When it all gets to you, and you reach for a glass of wine – why are there three old bottles of the white you don't like, and none of the red that you love?

Unless we are rich enough to have housekeepers and personal assistants, we all have these moments, some minor, some more problematic. We just don't have enough time and attention to keep on top of everything. This inefficiency does not just affect B2C. In businesses, frequently, contracts roll over past their good, negotiated terms and onto the rapacious default terms – because there is nobody around to watch and care. Or the service that could save our business a ton of money is overlooked, because it is nobody's job to find a replacement.

Every time we miss the opportunity to review or restock, a selling window is lost. The utility that the product could have provided at that moment is lost forever. If every customer were continuously vigilant, many of those moments would translate into sales. Scaled up – that would lead to a bigger market size overall.

> *"People don't simply buy products or services.*
> *They 'hire' them to make progress in specific circumstances."*[4]

So said one of the 21st century's most influential business thinkers, Clayton Christensen, in reference to his insight on customer "jobs to be done". If human customers obtain products to make progress in the jobs they want to be done – what impact would Machine Customers have on the situation?

One answer is that Machine Customers will be crafted to make progress on jobs that humans can't or won't pursue with sufficient capacity or vigor. Take office window cleaning, for example. How often are the windows on your office building cleaned? Because of the height of office blocks, the risk and cost of cleaning, most windows are quite a lot grimier than we would like. When you drove to work and stopped at the traffic light yesterday – did you notice the trash in the gutter and the branches overhanging the sidewalk? What about the burned-out streetlight and the broken fencing? There are just so many jobs to be done that we humans don't have the economic capacity to keep track of them. But machines will.

Gradually over the coming decade, more and more robots will be created, both physical and in software – to do the work we want to be done but cannot monitor continuously. Advances in predictive, generative AI, and agentic AI will enable each of these machines to also become a customer – for upgrades, for parts, for paint, for cleaning fluid and for working-at-height accident insurance. The jobs to be done are endless, and the insertion of these machines into the economy will increase the size of many markets.

Take the simple action of car cleaning. It consumes products like detergents, waxes, and cloths. You wish your car were cleaned more often, but the auto industry is already betting on a mobility future where fewer of us choose to own cars, and more of us use road mobility as a service instead. What will happen to car cleaning in a world full of autonomous vehicles operating in Uber-like ride-share services? Each customer will expect the interior and exterior of the robot vehicle to be clean. That vehicle might be cleaned not once a week but instead many times every day, like an airplane seat. That vehicle will be a customer of the cleaning services and the consumables it uses. All those robot vehicles together, acting as Machine Customers, could massively increase the size of those cleaning-related markets.

Another way Machine Customers will increase market size will be to spot and act on currently overlooked moments of need and opportunity. They will detect flaws, omissions, gaps, shortfalls, mistakes, and unexpected needs. Gartner has called these situations 'business moments' (a concept that Don helped to develop) - a passing window when a need arises and is addressable - but they are easily overlooked and lost.

Business Moment. A business moment is a transient opportunity where people, data, businesses, and things work together dynamically to create increased value.

Business moments happen all the time, such as when you...

.. wish you had brought an umbrella.

.. would kill for an ice cream.

.. still don't know how to get those garden shears sharpened.

.. would hold a meeting of your club tomorrow – if only you could find a conference room.

Machine Customers will make more business moments detectable and addressable. Their sensors will see situations and their connectivity will seek out market solutions and they will evaluate possibilities in real-time, using APIs and cloud-based marketplaces. Sometimes the resources that fill the detected need will be machines. Other times they will be humans. Gig working will often be part of the solution. For example, imagine you agree today to hold a get-together over dinner for a few friends next Saturday. You put it in your calendar and do not think much about it until Friday. Then you realize that though your recently remodeled kitchen is looking sweet, the oven is a bit gross. It's even been smoking a bit lately. But do you really want to spend a chunk of Friday evening scrubbing it clean? A cleaning service would be better – they have the steam wands, scrapers, and chemicals solutions that you do not - but it's too late to book them now – ugh!

In a Machine Customer world, this situation might play out differently. The connected smart oven would know when it's dirty. Its light sensors could easily see that the glass door is almost opaque with the burnt-on grease splatter from the recent roast duck incident. The intelligent assistant software agent that helps run your home has access to your calendar, and it 'knows' you have dinner for several people scheduled for Saturday. How hard would it be for the machines to organize a cleaning service to come in at a suitable time, via your smart lock front door, by posting the job to be done on a gig work site? And while the intelligent assistant is on the case – it might also order a new room fragrance and propose three different fresh-cook meal kit options, to make you look like a culinary superstar.

Winners and Losers in The Battle for Machine Customer Growth

Machine Customers will be even bigger than the earlier stages of digital transformation—representing not only a new sales channel, but a whole new category of customer. It's a profound shift that will challenge how you think about and grow your business.

Three Words Will Mark Out the Winners – Open, Rational and Scaling

Machine Customer markets must be open markets. We cannot emphasize this enough. The initial reflexive temptation to try and create and control a 'walled garden', a closed ecosystem of proprietary solutions for a select group of Machine Customers must be resisted. Those who try for proprietary control will lose out to those who create an open approach. There is a fundamental reason for this - human customers must trust the machine before they can delegate to the machine. You will be comfortable giving up purchase decision-making to a machine only if you believe that the machine is truly operating on your behalf, with your interests uppermost. For that to be true, the machine must be free to shop in an open market.

This situation isn't binary, of course. Between open and closed, there are many shades. Amazon operates a platform for many businesses to sell through. Apple operates an app store through which many developers distribute their software. Amazon and Apple sell their own wares on their platforms, but crucially they don't limit their platforms to only selling their own merchandise. Despite this obvious and important lesson, many businesses will still be tempted to try to create their own narrowly controlled, vertically integrated Machine Customer system. They will fail. This is one reason the smart kitchen appliance world is still a forever future vision – not an everyday reality. It's the 2020's already – where is my connected fridge that reorders chilled groceries? One big reason is not the lack of technology but rather the lack of an open platform. The big appliance makers have played with the idea; however, they have often attempted to implement it in a narrow and exclusionary way that has yet to give grocers the room they need to create a thriving ecosystem.

Machine Customers Will Be Logical and Rational - So Must You

The behaviors and capabilities that have brought success in the past have been honed and optimized to perfection. But all those skills are focused on what behavioral economists call the "bounded rationality" of human customers, who can be led to...

.. glance at a product that was photographed being worn by an attractive model.

.. burrow an "earworm" advertising jingle deep into their memory.

.. believe a "two for one" price is a better deal when they only need one.

.. mistake the size of the packaging for the volume of product inside it.

.. think that if they use a product, it will make them look younger.

.. sign an industrial contract because they had an enjoyable dinner.

We could add many more examples to that list. Humans are not entirely logical and rational. They don't always calculate well under time pressure. They can be swayed by quite vague associations. Companies optimize the way they design products, market them, and sell them based on these realities. But Machine Customers may not be fooled so easily.

So, what happens when Machine Customers start to invade a company's market? For B2C companies, it could be disastrous. As Aron Rajan, Global VP of Technology Consumer Digital Experience at Unilever, told us: "I think that when it comes to consumer brands, this is potentially the ultimate existential threat"[5]. It's good that he has seen this coming, giving his company ample time to adjust. In a worst-case scenario, another company may be so blind to the problem that it may not see the new Machine Customers at all. We saw this pattern in the early years of e-commerce. Initially, many companies assumed that online selling was a niche thing that could be safely ignored - even when it was growing fast. Some waited too long to catch up. Remember Blockbuster? In 2004, there were over 9,000 stores in the video chain. Today there is one, in Bend, Oregon. You can catch a documentary about it on Netflix.

Of course, we understand that digital disruption does not happen overnight. E-commerce took twenty years to kill companies, not two years. Machine Customer arrival will follow a similarly relentless but invidious path. Forget the "burning platform," a better metaphor is the fabled boiled frog. In the end, companies that jump out of the pot will be those that give up the presumption that proficiency in psychological influencing is the main key to future customer success. As leadership thinker Marshall Goldsmith likes to say, "What got you here won't get you there".[6]

Will Machine Customers impact a company's sales and marketing strategy and core competencies? Very likely. Machine Customers will be calculating, data-driven, and algorithmically deterministic. Their rationality will be simplistic at first, then more complex as they evolve and learn. The number of factors and parameters they compare before deciding will change and grow. So, a company that becomes good at inserting new decision factors into markets may do well. How Machine Customer neural networks learn from data examples will influence their behaviors. So, companies that work out how to feed them the right example datasets and influence their algorithms will gain the upper hand, as Machine Customers communicate with each other to cross-inform decisions and choices. Companies that analyze and understand such word-of-API interactions will be better placed to intersect and shepherd them.

At every turn, your company will be challenged to rethink its go-to-market approach based on logic, math, and data. Companies will ask themselves questions like:

- Why are we losing product feature comparison battles?
- Why do Machine Customer algorithms favor different pack sizes?
- How can we add more pictures for Machine Customer image recognition algorithms to absorb?

Some people are already thinking about how this might work. For example, Jeff Lewandowski, president of Andrew Reise Consulting, shared his perspective on marketing to an algorithm on Google's YouTube platform:

> *"There are a lot of individual producers of content on YouTube. Their role is to expand their overall footprint because of YouTube's revenue model. So, if I'm a producer of content and I'm able to acquire 5000 subscribers, I can then participate in the revenue share opportunity with any advertising that YouTube plugs into the videos. These content producers are motivated to add subscribers because they get paid on a per view basis, just like YouTube does. Part of what they attempt to do is influence their subscribers to train the YouTube*

algorithm. They say things in their videos like, 'Smash the "Like'"button' and if the 'Likes' get to a certain threshold, then YouTube will automatically suggest the video to new YouTube users. From the perspective of the content producer, developer, they're already marketing to a machine by asking their customers to like the video." [7]

Being able to scale up or down quickly will matter a lot. Winners will thrive on it, while losers will find it impossible. There are a few reasons why this will be true. Firstly, as Machine Customer markets emerge, they will tend to follow non-linear growth curves where the take-off point is hard to predict. An occasionally updated, pre-determined capacity planning approach will not fit well. Second is that Machine Customer markets will naturally settle on just a couple of platforms, so the race will tend to be won by the best scalers. The third is that once Machine Customers arrive in volume, they will tend to swarm, which will cause sudden up-swelling and down-ramping periods. One of the benefits of human customers is the diversity of action that varying shades of irrational behavior induce. In contrast, machines tend to all see things the same way at the same time using the same sort of logic. *It is unusually cold today. The odds that winter will arrive early are up 87%. Execute soup-buying subroutine.* This phenomenon has already been experienced in stock markets – the most advanced Machine Customer industry so far. Mechanisms have had to be devised to cope with the problem of 'flash crash' - when all the bots follow the same logic or trigger, and all dump an investment at the same time.

Machine Customers Will Eat Markets Silently

Machine Customer insertion into markets will follow the now classic non-linear curve that deceives early on with its apparent lack of progress, then hits a sudden take-off point. In one of our previous books – Digital to the Core[8], we observed this as a fundamental characteristic of modern digital business. The early years revenue curve of the now large and quite famous apparel company ASOS is a clear example. For almost the whole first decade of its existence, the company's modest revenue growth made it look like a hobbled zombie survivor of the dot com crash that, for some reason, wasn't quite dead yet. During its second decade, the company's revenue grew from under 200 million pounds to 3.5 billion once it perfected the art of selling clothes online profitably – something once thought to be almost impossible.

Technology-related changes in business have a habit of creeping along for a long time until a tipping point is reached - and then taking off like a rocket. When three things fall into place – technology performance, market culture, and industry regulation - then the business will suddenly ignite. The companies involved understand they are working ahead of the right market conditions and technological progress. This effect is partly amplified by the speed of sharing (social media and the like) that the internet has unleashed. As soon as the innovation really works, everybody is only a few steps away from somebody who has experienced it – and they very quickly hear about it and want to climb aboard.

These slow starts can lull sleepy incumbents into a false sense of security. Their competitor intelligence units spot the market innovation early on, analyze it and keep track of it. They may even decide to run a business experiment that copies it, but that test seems to show it is not a promising idea. They watch the upstart for a couple more years, then ignore it. The idea was silly, and the protagonists did not understand the industry – they told themselves. They declare it a dead duck. When it starts to rise a couple years later, they don't notice it. Or worse, they lack the humility to realize they were wrong and deliberately suppress conversation about it, and in another year or two, have the wrong kind of Kodak moment.

In late 2022, a very public Custobot incident at Ticketmaster brought the market power of Machine Customers into the public spotlight. Global music superstar Taylor Swift released tickets to her latest US concert tour through the site. However, many fans had difficulty accessing tickets and the resulting complaints led to Ticketmaster issuing an apology. The statement on the ticket giant's site explained:

> *"Keeping bots out of queues and avoiding overcrowding helps to make wait times shorter and online sales smoother."*

But also

> *"..this time the staggering number of bot attacks as well as fans who didn't have codes drove unprecedented traffic on our site."*[9]

Ticketbots are used by 'scalpers' – gray market ticket resellers – but they are also used by individual customers. This incident shows just how hard it can be to know or predict the level of impact Machine Customer insurgency will have on a market. In some cases, it might be hard for an organization to distinguish human customers from bots, as Ticketmaster found out in early 2025 when it sold tickets to an Oasis concert in the UK. They though some human customers were bots and mistakenly cancelled their tickets. [10]

The arrival of Machine Customers in markets will follow a common pattern. There will be a flurry of early excitement followed by skeptical comments of wise old industry insiders. During the stealth period that follows, the introducers of the machine-customer future will work out all the kinks until they get the recipe exactly right. Then it will steal the market at an extremely high pace. Revenue or inventory may fall suddenly in almost any kind of industry. For example, in the UK shortages of driver's license test appointments have been caused by resellers equipped with shopping bots making multiple appointments they could then scalp.[11]

Doubling down on their human market and perceived higher value human customer service capabilities, the losers will find their cost of sale gradually increasing even as their revenue and total addressable market appears to shrink. It will be incredibly tempting to say that the only customer that really matters is the human one[2]. That belief will be repeated by some managers even as more and more buying decisions are gradually assumed by Machine Customers that they are unequipped to persuade. It will be hard for those inside the corporation who have been the stars of human-style marketing and selling to cede their positions of power and privilege. Worse still, the instinct for self-preservation may cause them to turn their capabilities in human manipulation on their own corporate cultures.

One executive told us that if the Machine Customer future does ride roughshod over its competency at brand marketing, the company would probably pivot to rely more on emerging markets. This assumes Machine Customers will arise first in advanced wealthy economies, while the less advanced remain human-customer dominated. That is a reasonable defensive strategy – for a while. It might protect a large company for twenty years or more. But you can only buy so much time with moves like that. Eventually, the technological future will come for you, just as it did for Sears and Blockbuster.

---

[2] In the late 1990s one of the authors was involved in a heated exchange, with a marketing strategist at a large airline. The executive stated that there could never be such a thing as online customer service. For them, the concept of service was, by definition, human. However today we all check in for a flight online without a second thought. If the technology fails and we are forced to go to a human check-in agent, we are irritated – no matter how warmly they greet us.

Of course, it's also possible that there will be no sheltered backwater. Sometimes emerging markets leap ahead of more mature economies. Examples like Alipay in China and M-Pesa in Africa spring to mind.

Machine Customer Losers Won't Even Realize They Are Being Played

Sometimes, incumbents may not even realize what is happening.

At a conference in May 2018, Google CEO Sundar Pichai demonstrated an AI technology so advanced it made his audience gasp and laugh nervously at what they were hearing and the implications. The demonstration was called Duplex. You can watch it on YouTube[12] , and we encourage you to do so. An AI assistant called up local businesses by phone with voice synthesis and conversational capabilities, so lifelike, that the customer service people being called did not know it was a machine. It called and made reservations. It sounded so real - it literally 'ummed and awed' as it pondered date options - that the people picking up the phone did not know they were speaking with a Machine Customer.

In the following days, there was a media backlash against the idea. People weren't ready yet, so Google quietly shelved it. However, the technology will not go away, it is just biding its time until we are all more ready. Synthetic voice actor technology indistinguishable from a human is already commercially available. The technology has resurfaced again with Eva – a software platform that automates patient scheduling for doctors' offices, and even contacting patients directly. Google Labs is testing "Ask for Me," a service where instead of calling a company yourself, you can let Google's AI do it for you, handling tasks like finding the best price and checking product availability.[13] Generative AI is accelerating this kind of Machine Customer capability. For example, a small but impactful company called DO NOT PAY has developed a ChatGPT-enabled Custobot that can handle such complex transactions as successfully renegotiating an internet service contract with Comcast.14It was so convincing that a human chat customer service agent at the telco offered it a $150 discount.

The losers in the Machine Customer wars ahead will continue to assume they are dealing with a less than rational human. But the machine will see through them, and in negotiations, its carefully calculated and hyper-logical choices will eat up their margins. It will see through their carefully fogged psychological smokescreen to find the best value. If the seller's go-to-market strategy assumes humans mostly make irrational and lower-value choices, then the seller will start losing the game more often. One cold-call thwarting service provider – the Jolly

Roger Telephone Co., offers a generative AI GPT-powered phone-answering voice chatbot that strikes up conversations with telemarketers. It is designed to waste their time convincingly and deliberately.[15] Daisy, the AI-powered granny created by Virgin Media's 02, uses AI to make life harder for phone scammers and protect potential victims. Daisy tied up scammers for more than an hour, leaving them frustrated and empty-handed.[16]

Daisy was created as part of an initiative to combat phone scams targeting older adults. Designed with advanced natural language processing and a touch of humor, Daisy mimics the speech patterns and demeanor of a kind, slightly forgetful elderly woman. Her charm lies in her ability to keep scammers on the line while giving them absolutely nothing. Popular scam-hunter Jim Browning, whose YouTube channel boasts 4.4 million followers, contributed his expertise to Daisy's development. Sometimes sellers won't even realize 'who' they are playing against.

The Decisive Battlegrounds of Machine Customer Markets

Speed will be a critical factor in securing Machine Customer markets. Machines can often operate at a far higher speed than humans. Sales cycles that take days for human customers could take nanoseconds for Machine Customers. This could be a dangerous mismatch for suppliers used to having a relatively long time to come up with an offer:

- Even today, in quite a few specialty B2B situations providing a price still involves a human looking up, calculating, or at least checking a price before it goes out to a client.
- In many B2C markets – even those substantially online such as insurance – the full range of prices and coverage options can only be accessed by phoning a person who then must go through lookups, processes, and calculations, and perhaps supervisor approvals.
- Simple facts about products, such as technical specifications, can only be found by complex searches within documents intended to be read by human eyes.
- Today's interface to the market is often hostile to machines. We have all experienced being challenged to click on pictures to prove "I am not a robot" before being allowed to enter a website to buy something.

If it is common for your company to say, "we will get back to you on that later today" then you will likely be an irrelevant dinosaur to a Machine Customer. To sell in their world, we will have to devise new and almost instantaneous methods of providing answers.

Imagine for example, the AI assistant that helps manage a consumer's home and needs to find a window cleaning service. First, it must identify services online that might be willing to bid for the work. That will probably take place via an online services marketplace. The only players will be those who have API connections to the cloud-based market. That won't exclude individual gig workers – but they must be connected by an app to play at all.

Then there will be a price negotiation. How much does it cost to clean this home's windows? The responding service provider must calculate a price quickly enough to stay in the running for the job. Sending a person in a truck to size up the property won't cut it. What might work is having an algorithm that can analyze the Google Street view image of the front of the property using image recognition. It could then infer the probable complement number of windows on the back of the property and check the known number of windows on other properties nearby that it has cleaned in the past. It would then guesstimate the square footage of glass and the complexity of the job, compare that to other jobs' cost and come up with a price, all in a second or two.

None of that is bleeding-edge in technical terms. It could be coded by stringing together widely available API functions from a variety of cloud service providers. The point is simply that it is not the way the window cleaning industry wins customers today. The slow pricing responses of today rely on the far higher delay tolerance of human customers.

Product Comparison Will Be a Constant Competitive Battleground

Product feature comparisons will likely be more important in the Machine Customer world.

Think about any significant purchase you made recently – a bike, a vacuum cleaner, a small fleet of a dozen backhoes. How many product features did you compare in your search, and how rigorous was the trade-off between them? Maybe you are someone who creates a spreadsheet and geeks out a bit. Most people don't. You might have compared five key factors or maybe ten, if you're obsessive. But you didn't compare 50. This is exactly the kind of customer work that humans quickly tire of. But machines are relentlessly good at it.

Machines can and will trade-off harder factors (size, weight, battery energy density, tensile strength), far more easily than humans. However, some of the softer and more subjective factors could be sidelined – at least initially. There may not yet be good, standardized ways to communicate the mouthfeel of different potato chip shapes or the perceived differences in what it "says about you" to be wearing Prada vs. Louis Vuitton.

This sets up a significant future battleground. Companies will compete to inject new comparison factors into the equation and to find new ways of describing those factors. These will include soft factors gleaned from social media analytics and biometric sentiment analysis. There will be a constant need to inject fresh differentiating factors to disrupt and slow down the commoditization effects of machine comparison.

An Ability to Cope with Unpredictable Volatility Will Be a Key Competitive Factor

In many industries today, a great deal of work takes place to predict demand patterns based on seasonality and other factors. The summer flying schedule for an airline is different from winter. The production of lawnmowers and leaf blowers ramps up for summer and for fall. The occasional crisis, such as a hurricane or a pandemic, may cause planning chaos, but those episodes are exceptions. In Machine Customer markets, sudden inexplicable swings might be more common.

Machine Customers will tend to herd and swarm, sometimes causing wild volatility. As we mentioned earlier, this is often seen in stock markets – one market where most buyers these days are already not human traders but algorithms. In our research interviews for this book Tim Quast, a veteran investor-relations professional and president and founder of ModernIR, told us candidly:

> *"By our measures 50 to 55% of trading volume is done by high-frequency traders, which largely means machines because it's too fast for humans to be able to participate."*[17]

Quast also said:

> *"It is being done by machines that don't understand anything about the company whose stock is being traded. The algorithm doesn't care about the stock. Often it doesn't even care about the news about the stock. There are some algorithms that*

> *analyze news, releases, analyst reports, and that sort of thing. But at this point, more than half the volume is simply programmed – if it's going down, try to sell now, I can buy it back a little bit lower later – which is just momentum."*

There is evidence that AI-enabled algorithmic trading machines are getting more sophisticated and delivering better results.[18] However, the kind of swarming momentum effect Quast described is likely to be a big feature of Machine Customer markets, especially in their early years. Companies will need to develop sophisticated methods to handle this crazy side of the hyper-rational. Beyond the simple 'circuit breakers' that stock exchanges introduced, companies may need to consider randomization to help dampen the effect of algorithms all making the same calculation at the same time. At a higher level, CEOs will need to think carefully about business models and operating model designs that thrive on exploiting volatility along the lines of thought suggested by Nassim Nicholas Taleb in his book *Antifragile*.[19]

What Your Company Must Do to Benefit from Machine Customers

To gain control over the machine-customer future, you need to do three things: make your own Machine Customers, find a path to those that will control your markets, and go after the growth curve the moment the transition to custobot control begins.

Make Your Own Machine Customers If You Can

For some businesses, making their own Machine Customers will be a real possibility. Cars will become customers and so will capsule coffee machines. It won't surprise anyone if GM Cruise vehicles start buying fuel, tires and GM-manufactured spare parts – all with transaction fees that go back to GM. It won't surprise anyone if connected Nespresso machines integrate capsule counting with the subscription model to essentially become the fully delegated buyers for their human owners. It might be tempting to think that huge companies are the only ones able to create and control their own Machine Customers, but that's not the case, as the story of one not-so-vast Swedish washroom products company proves.

Essity had its origins in the paper and pulp industry company SCA. Over time, it expanded from supplying paper hand towels to supplying a wide variety of washroom and hygiene products and was spun out from its parent in 2017. Since around 2015, it has offered a digitally enabled system of washroom products called Tork Vision Cleaning[20]. In the washroom of a hotel or hospital, you will see the boxes on the wall that dispense liquid soap and paper towels. The internal electronic modules are not complex. Inside the dark box, a simple detector can sense when the paper towel stack is nearly finished and send a Wi-Fi signal to janitorial staff to restock. It's a simple, well-executed idea. A smart high school kid with a Raspberry Pi could probably build a reasonable prototype of such a thing over a weekend. The key insight here is that it does not take much investment to start experimenting with your machine-customer future. Once the prototype works, move on to a professional reference design for the product. Take that to contract manufacturers in China via Alibaba and move on from there.

So why is Tork Vision Cleaning not already a renowned international MBA case study? Remember, trends like this start small with modest growth and then take off later. Tork Vision Cleaning is a reasonable success, but the company doesn't shout about it, probably because the sales results so far are modest in comparison to its traditional core business lines. A B2B industry like washroom services doesn't change its operating culture overnight. However, one industry specialist analyst forecasts a half billion dollars connected washroom product market by mid-decade.[21]

Another important point to note about the Essity example is that the starting point was an inert product. The washroom paper product dispenser wasn't originally electronically controlled or even electrical. But many more things are already becoming so these days. For example, US company Herman Miller now offers a collection of connected office furniture - desks and chairs under the system brand name 'Live Platform'[22]. Canadian company BigBelly[23] makes an outdoor connected and signaling trash can system that is in use in the streets and parks of major cities all over the world. These kinds of devices are only capable of basic things like telling you when they are ready to be emptied, but digital and connected products will get a whole lot smarter quickly. From there, a more sophisticated Machine Customer category can emerge.

Take vacuum cleaners. Robot vacs like those from Samsung and iRobot now make maps of the interiors of the homes they are serving. They can 'see' changes to furniture positions and locations. The maps they create live in the cloud and can be shared back to you as a service[24]. It's easy to imagine a future where the cleaning bot could sell its data in exchange for vacuum bags or even deals on new rugs. Figure 6 shows just how detailed these maps can be. Maybe the value of this information is one reason Amazon tried unsuccessfully to acquire iRobot in 2022.[25]

Figure 6 - Home Room Map Entirely Generated by A Robot Vacuum Cleaner

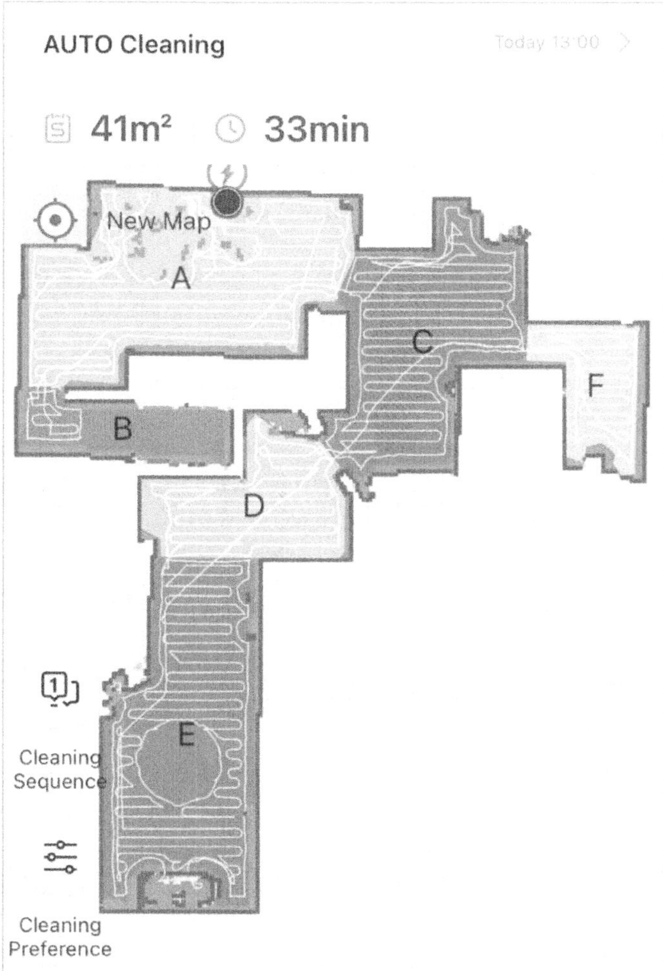

What customer jobs-to-be-done might machines like this become able to take on eventually, with this kind of additional sensing capability? Already, the Dyson v15 Detect[26] handheld vacuum cleaner can measure and report on the different sizes and categories of dust particles it picks up via its laser sensor - information that someone, somewhere might well find valuable.

If your company already has a large market share, it may have a chance to turn its *existing* products into smart Machine Customer systems. One of the first lines of business will be machine servicing and replenishment of consumables – just as HP has with Instant Ink. However, if your company is small, perhaps the introduction of an innovative new Machine Customer could be a way to grow fast.

Find a Path to the Machine Customers That Will Control Your Markets

If you can't make your own customers (or even if you can), it will be crucial to methodically identify all the possible "things" in your industry that could become customers. The questions you should ask are simple enough:

- What are the activities in the life of your customer (user, not just buyer) in the context of using or consuming your product?
- What objects or machines are related to those activities?
- Could those things become Machine Customers?

For example, a big sausage company might say barbecuing in response to the first question, fridge, firepit, and meat thermometer in response to the second question, and "yes – possibly" in response to the third.

So already, they have sketched out some of the future machines that *might* one day become customers. Perhaps they should speak to Samsung, Electrolux, and other appliance makers about the state-of-the-art innovations for those connected devices. If they hear that Weber (a famous maker of barbecues) is adding smart digital features to its outdoor grill – then maybe, they should speak with their R&D function too.

This exploration can get strange and confusing, especially if starts in the wrong place. For example, a tire company might struggle to see how its tires could become customers. What could a tire really 'want' or shop for on its own behalf or for its owner? That line of thought doesn't seem to make much logical sense. However, it seems very likely that cars will eventually become customers for tires.

Once you identify a few promising Machine Customer candidates, start sketching the design of the services they will need. What end-consumer needs will Machine Customers be capable of detecting? How does the Machine Customer journey compare to the human customer journey? How would that machine do its search and comparison shopping, and how could you best intersect that to capture its order? Words like platform, marketplace, ecosystem, and API start to appear in the conversation from there.

Go After the Growth Curve the Moment You Detect It – Do Not Wait

It is tempting to see this whole idea all as some far-off future thing. Why bother doing anything yet? You have no doubt read or heard it said many times that being a "fast follower" is the way to succeed. After all, Microsoft didn't invent the spreadsheet, and Apple didn't invent the MP3 player - right? Well, things have changed a lot since the 1980s and 1990s. Plenty of successful companies such as Amazon, YouTube, Tesla, and others did start out as first-to-market originators. What's more, they have attained very powerful oligopoly positions. Technology-enabled markets often seem to end up with just two or three dominant players.

Machine Customer growth will be like that of most tech markets – it will follow exponential curves that converge into an oligopoly. Progress moves quickly once the rate of innovation is at least partly driven by Moore's Law effects. We have seen this as products become digitally enhanced. Consider recent progress rates in the electro-digitalization of the car or the substitution of cigarettes by vapes, or how much the concept of movie and TV consumption has changed in a world of mobile devices and streaming services. If you don't get in early, you may not get in at all.

Once your customers become machines powered by Moore's Law, you should expect similarly rapid rates of evolution. Nonlinear progress is a hard act to follow. Imagine if the first company to initiate a Machine Customer approach to a market acquires 5% of the total revenue. Within a year, that could be 10%, and within another year, it could be 20%. They probably picked off the easiest sections of the market – where human customers were most ready to cede control to machines.

What happens when you start your copy and catch-up move? You could go head-to-head and try to replace their Machine Customers with yours. But you would have to talk to the owners, not the machines, and the owners have already stopped being interested in being active human customers. Persuading – or even accessing those human end-users would be hard and costly.

Alternatively, you could go for the 80% of the market that is not yet Machine Customer-controlled. This 80% was less amenable initially, but perhaps you are lucky, and by this time, they are ready to engage. So, you take 20% over two years – the same penetration rate as the pioneers. But the pioneers didn't stop and wait for you. Over those same two years, they took a further 40% on top of their initial 20%. So, four years after the start – they are at 60, and you are at 20. The point is this – you must be a very fast follower to stand any chance of even being second in tech-centric markets. Between the two first players, there is only 20% left for everybody else – and that remaining market might be a very change-resistant group.

This means your company has only two viable choices – either pioneer your market's Machine Customer future or be prepared to be an extremely fast follower To be warmed up and ready to go, will require your company to have created a sophisticated and highly capable human-customer digital business platform already. The many human UI channels will be backed by a sophisticated microservices API platform. There will be digital product management, agile software capability for products, and mass agility for the platform. Advanced data science will be applied for feature comparison, dynamic pricing, and negotiating position. The list goes on.

Conclusion

Machine Customers will be part of your business. They will be everyone's business. Whether you are selling insurance, payment processing, tires or soap, the nature of your competition will soon be changing. Machine Customers will bring tremendous growth opportunities and may increase the total size of some markets – while shrinking others. Either way, they will be an unavoidable force.

Key Takeaways

Machine Customers will create growth opportunity.
Humans are often not good customers. They are inherently distracted. It costs a lot of money to win their attention and even more to keep it. Machine Customers, on the other hand, start focused and stay focused. While their ruthless efficiency may strike hard bargains - they won't forget when it is time to shop and replace. They won't miss any of the fleeting recognitions of need we call business moments. In some situations, their attention will result in the growth of the overall market.

Action: Start making your own Machine Customers or start serving the ones who appear early in your markets

When it comes – the battle for Machine Customers could be brutal.
There will be winner-takes-all situations because tech markets tend to gravitate quickly to the best platform solutions. The rate of change will be non-linear, and that hockey-stick market growth curve will catch people out. If you are to engage and win your share in the battle for Machine Customers, you will need to change some of your deepest and most revered core competencies. A focus on functional vs. emotional advantages in your go-to-market will be essential. Above all you must substitute logical market algorithm capabilities for skills in human persuasion.

Action: Invest deeply in data science, AI, algorithms, and an open API platform.

Machine Customers will often arrive silently.
Don't believe you can catch up later. It is better to start small now so that you are rehearsed and ready. Think and act like a start-up.

Action: Identify what machines could become important customers in your markets. Start with minimum viable versions and keep on experimenting until you get it right.

Chapter 5

Machine Customers Are Made, Not Born – Who'll Be the Makers?

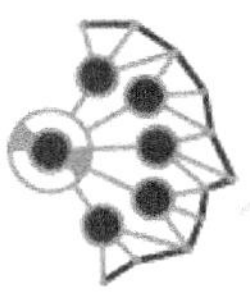

As Victor Hugo said: "On résiste à l'invasion des armées; on ne résiste pas à l'invasion des idées"[1]
(yes, of course we do languages)

The masters and makers of the Machine Customer universe are already emerging and taking key positions

Build a Trusted Custobot, not an Invading Army

Major markets can only arise when three key elements are in place: a mechanism to create trust between parties, a scaled marketplace or platform and a regulator to assure safety and competition.

Let's Start with Trust Because Nothing Will Move Without It

History tells us that delegating trust to a machine does not always come quickly and easily. A good test of trust is money – we all worry about it and have a natural protective instinct towards it. We don't give up control easily. So, consider the

---

[3] Victor Hugo's original French quotation is often loosely paraphrased into English as: "Nothing is more powerful than an idea whose time has come."

challenge facing the inventors of the early automatic teller machines. Patented in the early 1960s, it took quite a while before the concept moved into the real-world. Today we all accept that a cash dispensing machine is highly accurate and the most convenient way to obtain banknotes, but exceptional circumstances were required to nudge people into using them. In London, a 1967 banking labor dispute over Saturday working caused the banks to push out the new technology. They also brought in a popular TV sitcom star to try it first, in a very public press photo. A decade later, in the winter of 1977, widespread snowstorms in the U.S. disrupted bank operations and spurred consumer adoption.[27]

Trust in machines doesn't come free, it must be earned. We are persuaded by others whom we trust (a TV star), or if we are forced to try the new device in a crisis (a snowstorm). Andrea Cicollini, CIO of Amplifon, the world's largest retailer of hearing aids, believes getting people to trust smart machines will be difficult:

> *"When there are no real regulations on the machine learning framework, what boundaries do you put on how smart and what type of decision my machine will take for you? As of now, there is no legislation, and there are no boundaries; there is just "go for it." We need to set up a consortium that sets up the rules. Otherwise, if this goes out of hand, I believe it's going to be then difficult to go back."*[28]

If you are over 40 years old, you might remember how you felt the first time you used your credit card to pay for something online. Back around the millennium, when e-commerce was new, people were quite uncomfortable and mistrustful. Many swore they would never shop that way. Twenty years later, in 2021, Amazon was the fifth most trusted brand globally and Google the number one - as measured by Morning Consult[29]. It's worth noting that Forbes ranked those companies in similar positions by brand value in 2020.[30]

If technology brands have risen to such lofty heights in the minds of consumers, we can conclude two things. First – there is a good chance that consumers will be prepared to trust Machine Customers because they already trust the tech sector. Second, companies such as Amazon and Google have a head start. For example, Amazon already has a service that can detect counterfeit goods using AI.[31] It is likely this will extend to counterfeit AI bots. But there are other kinds of companies that have a strong trust hand to play too. For example, Nike, Nestle, and Colgate-Palmolive are right up there in the top trust rankings. For Machine Customers operating in category contexts such as athleisure, nutrition, or hygiene, the tech giants don't necessarily hold all the cards.

But prior trust will not be enough. People will only delegate their customer work to a machine if they are confident that the custobot:

- Operates on their behalf
- Can make free choices
- Can be relied on

Let's think about Amazon's Alexa for a moment. Do I want to delegate some of my shopping work to it? I know it doesn't really belong to me, and I didn't code it. In fact, its algorithms are opaque, even to experts. The situation is a bit like working with a financial advisor who is selecting a mutual fund portfolio for you. The adviser and Alexa are both part of a sales system. They are making a margin on what is being bought for you. That source of income could bias their choices. Perhaps they are tied to only a few products and never consider others. Bias in Machine Customers is something we will have to deal with as far as humans are setting the rules and programming. Just as we are on guard against biased salespeople, we must also be on guard against biased custobots, especially those whose AI is trained on a dubious dataset.

We will need reassurances that the Machine Customer is working substantially, if not wholly, for us. The Machine Customer must be operated and presented as personal support, not a sales agent in disguise. Some of this will depend on being psychologically comforting, and some on finding trustworthy business partners. Amazon's Alexa doesn't belong to you, but it probably *feels* like it does. It uses its familiar voice in your home every day, and that voice comes from a smart speaker device that you bought and own. This kind of trust-building is psychological comfort, but you will still need evidence that Alexa is really working for you.

That trust in the Machine Customer will be built from multiple elements. First, the machines will need to display overtly customer-centric behavior to counter our suspicions of them. One obvious way will be for the machine to switch providers from time to time. Whether it's for price or for product performance, the Machine Customer should regularly delight us by buying better and letting us know that it did, or by proactively switching providers to save money.

My Custobot Will Call Your Custobot

Another way Machine Customers will build trust is via peer affirmation and reassurances from friends, family, and wider social groups. I might trust my custobot more if I know it is linking to and relying on information from your custobot. A custobot operating in isolation isn't demonstrating the kind of group research and confirmation effects that human shoppers rely on. If we see some form of social collaboration between custobots that mimics what we do as human customers, that behavior pattern will reassure us. Evidence of social collaboration will need to go beyond the anonymized customer ratings of eCommerce websites, which are deeply mistrusted. The custobot will need to have some preferred, confidential, and intimate connections – a truly trusted sub-network that might include the custobots of your family and closest friends. We can imagine subnetworks of connected home virtual assistants exchanging information on which dishwasher tablets *really* work best in our hard tap water neighborhood.

Open market custobots will beat walled garden custobots in the long run because we will trust the former more. Amazon's Dash Replenishment Services created a platform to theoretically enable our trust because it hosts so many independent businesses. However, the Machine Customer device will also need to permit a range of buying possibilities. Today an HP Instant Ink printer *can* shop via Amazon Dash replenishment, and that's a start. So far, however, an HP printer acting as a Machine Customer must order cartridges from its manufacturer – and that's true for Brother, Canon, Xerox, and Epson too. Third-party ink supplies are not permitted. "At this time only genuine ink or toner produced by your printer manufacturer is available through Dash Replenishment Service" said the Amazon FAQ in mid-2022.[32]

Here is an example of the open market dilemma. If corporations insist on these kinds of restricted product relationships, many human customers won't delegate control. The convenience of Machine Customers is a benefit - but we will not yield control if we do not trust that the Machine Customer path gives us *at least* the same shopping rights. The market for third-party ink supplies is massive and important. Why would human customers give up that choice for the privilege of paying more? We believe that so long as companies contractually insist on tied relationships, Machine Customer markets will be hampered. Only open platform thinking will unleash the full power of Machine Customer markets.

Transparency will be crucial. As Aaron Rajan, Global VP of Consumer Digital Experience at Unilever told us:

> *"If our lives are increasingly going to be influenced by algorithms, every relevant consumer decision made by an algorithm on your behalf will need to be Accessible, Bias-free, and Comprehendible (ABC) – there shouldn't be any black boxes.*

> *History has taught us that even with good intentions – we can unintentionally create flawed systems; sharing the code (much like the open-source community has done for decades) will promote trust and self-regulation."*[33]

We will also want legal assurances that we are protected if things go wrong. What if a system is covertly tied to a single provider in a way that cheats us of our consumer rights? Where will the redress be? If our Machine Customer buys the wrong thing or suffers from a glitch and buys 200 jars of mustard when we only need one, who will pay? Insurance companies will have an opportunity to help indemnify custobot creators against unintended harmful effects of their programming. Lloyd's of London, a major insurance market, is addressing the growing risk landscape related to artificial intelligence (AI) and has introduced insurance products that cover liabilities arising from AI-related malfunctions, including chatbot errors and hallucination.[34]

Finally, regulation will be necessary to ensure the smooth operation of Machine Customer markets. It will be in the self-interest of smart governments to move early, simply because higher economic growth will be generated by Machine Customer activity. The detailed 'activity log' recording of the Web 3.0 programmable economy will make it easier for government authorities to track sales activity and tax it effectively. So, governments will have a double incentive to regulate early: tax base growth and taxation compliance. However, we have no doubt that Machine Customer systems will arise in many places *before* that regulation is framed. As usual with technological advances, we will see the dark side arise before action is taken.

Platforms Will Matter Because Scale and Openness Are Vital

Machine Customers won't ever be Boston Dynamics Atlas-style bipedal robots stomping around shopping malls, basket in claw. They will be substantially virtual and operate via cloud-based software API connections to electronic platform marketplaces. Other machines will deliver the purchase. In many ways, the human e-commerce ecosystems of today have built the basic infrastructure from which the Machine Customer markets of tomorrow will arise.

Already a vast electronic ecosystem makes it possible for you to order from wherever you are on a mobile device and have what you need delivered to your device (Spotify), your location (Uber Eats), or to a family member (1800FLOWERS). The software ordering is via API to an e-commerce service that triggers other API-enabled services and actions. For example, your Kroger grocery order might be picked and packed in warehouses operating Ocado's robotic systems.[35] Your e-commerce delivery will be routed by an algorithm. You can see how many stops away the truck is on a map – maybe one generated by Here, a mapping consortium company co-owned by German automobile manufacturers. Perhaps the delivery vehicle is electric and partly autonomous. Or the delivery vehicle may even be a fully autonomous robot, like a Starship Technologies ground drone or a Manna Drone Delivery aerial drone – both of which are already in limited operation.

These delivery platforms run in the cloud, connect with APIs, and have achieved huge scale through consolidation and standardization. That's because these platforms are marketplaces where network effects add value. The more nodes in the network, the higher the utility. The more users there are on WhatsApp, professionals on LinkedIn, or traders selling through Alibaba – the more useful it is to you.

From stock exchanges to eBay, buyers gravitate to where they can aggregate, compare prices and source most widely and efficiently. Scale brings more and more producers to the party. Such platforms are where the Machine Customers will flock, and where the biggest battles for control will take place. Companies such as Alibaba and Amazon have a very big head start. But that does not mean they have won already. The history of tech-enabled business change tells us that the initially dominant players do not always retain the crown – think of Compaq, Myspace, Nokia, and BlackBerry. As we transition from a human customer economy to a Machine Customer economy, some of the newer commerce tech arrivals may take the lead. Whether that is Shopify, Mercado Libre, Jumia, or another site being dreamed up right now in a dorm, there is room for new major platform players to arise. We think this is particularly likely in product and service categories that do not yet have highly consolidated, global human e-commerce platforms, such as:

- Insurance
- Food
- Fashion
- Furniture
- Business services

The platforms that win these markets will have the right mix of data formatted in the right way, as well as speed, low latency, reliability, security, and data privacy.

Also, crucial (and often underappreciated) will be developer friendliness. How easy is it for Machine Customer developers to learn and use the features and functions that the platform's APIs make available? How well does the platform provider listen and respond to developers and nurture that crucial community?

Will Machine Customers pay for access to the platform directly? Probably not – or not much, because keeping such fees low will be key to the rapid growth that leads to competitive success. Instead, the platform will make its money on secondary value streams, the way search engines and e-commerce sites have. What kinds of secondary value flows might arise in Machine Customer markets? It won't be easy to insert ads as a monetization method, but we can foresee several other value stream possibilities:

- Insurance. Because machines can go rogue and do unintended things, there will be an opportunity for insurers to provide cover that feeds off the transaction flow.
- Finance providers. As Machine Customers evolve, so will their separation from their human owners. Over time, Machine Customers will be granted more autonomy over the funds that they accrue and spend.
- Machine commerce optimization services. Loosely akin to the SEO specialists in human online marketing. These analytical companies will help vendors improve the visibility of their offers to Machine Customers.

Beyond the current digital power players, who else might have a chance at these markets? One group may be new digital companies that serve Machine Customer needs that the human-centric customer platforms have left exposed. An example could be cryptocurrency platforms. The arrival of Machine Customers is broadly contemporaneous with the rise of crypto. Some Machine Customers might 'prefer' the new blockchain-based currencies for a variety of reasons. That could create opportunity for companies such as Binance, Coinbase, or Kraken - if they survive the so-called 'crypto-winter'.

By the second quarter of 2025, major payment providers like Visa, Mastercard, Paypal, and Stripe announced plans to support payments from AI agents, in what is being called "agentic commerce" or what we might call "Machine Commerce."[36] These important announcements set the stage for a substantial increase in digital commerce transacted by AI agents acting as Machine

Customers. Also of interest are organizations like Tether.io and Nevermined who are creating platforms where AI agents can pay AI Agents.[37] Don Gossen, the CEO of Nevermined told us, "If Phase 1 of machine customers is a printer buying ink, Phase 3 is a constellation of AI agents buying, selling, and reconciling thousands of microservices a second."[38]

Other opportunities will arise in specific complex and large-scale industry contexts. For example, will VW or Ford try to create a platform on which car Machine Customers buy fuel, rideshare passenger drinks, and tires? However, the temptation to build a walled garden may limit the appeal of their platform, in which case, cross-industry system players like Here Technologies may stand a better chance because they can create market spaces accessible to many different companies. Nespresso's enlightened self-interest is an interesting example here. Its capsule coffee maker system was once patent protected, but when that period ended, many companies began to make and sell coffee pods for Nespresso machines. It has since struck partnership agreements, for example, with Starbucks, and took part in creating a pod recycling scheme (Podback) that is open for all capsules – including competitors. As capsule coffee machines have become internet-connected, they are already close to becoming Machine Customers that reorder capsules themselves. How much poorer would Apple be if its App Store wasn't open to other software houses? Nespresso's recognition that an open beverage-related Machine Customer market will be more valuable than a walled one, makes it more likely to become a beverage platform giant.

Progressive Regulators Will Be Builders Not Bureaucrats

Everyone who is irritated by regulation would hate living in a world without it. Machine Customer markets will be small and mistrusted without proper guardrails.

As noted earlier, functional Machine Customer markets will require collision avoidance controls. Machines that all do the same thing at the same time can be very dangerous. In financial markets, you get flash crashes. In swarming autonomous car markets, you might get gridlock and physical crashes. Neither would encourage humans to give up control and ask machines to do customer work for them. If there is any hint that Machine Customers are swarming and 'ganging up' to act in concert and corrupt markets, we will avoid them. The Machine Customer economy must be rules-based and well governed.

For Machine Customer markets to grow and thrive, there will have to be rulemaking, rule taking, and bad actor punishment. Business leaders often profess frustration with regulatory red tape, but the other side of the coin is that they really need some of it. Without rules that frame the market space and behaviors within it, it can be impossible to operate at all. We have seen rule-makers play a critical role in the history of tech-enabled business change. For example, until telecoms regulators stepped in to help allocate mobile spectrum and force interoperability between providers, text messaging didn't really take off.

Today, national aviation and transportation agency regulators are competing to create safe spaces for the development of aerial drones and autonomous cars. They know that national economic futures may depend on the rate at which they can shape and define orderly, risk-managed, and commercially dynamic rules for autonomous transportation.

In a similar way, progressive regulators will proactively assist in creating Machine Customer markets that are safe and fair to every stakeholder.

Enter The Corporate Directors of Machine Customer Strategies

Imagine the CEO of a growing German or Canadian company decides to open a new office in Beijing and start selling in China. That's an important strategic move. Few CEOs would try it without notifying the board and discussing it first. The same will be even more true as the company develops its Machine Customer market. The first challenge will be getting the board to accept that a machine can be a customer at all. As they wrap their heads around the idea, the board members will realize there are still big unknowns to solve.

Strategy, Opportunity, Threat Assessment, Corporate Values and Company Purpose May All Have to Change

As Machine Customers emerge, the issues thrown up will be material to the company's future and potential. Investor and stakeholder interests will be impacted. If the company pursues Machine Customers, will it grow or shrink the company's total addressable market? Depending on the product category and situation, it could go either way. When we serve Machine Customers, what is the cost of the sale? That might go down. But are machines better negotiators than humans? If so, the potential gross margin may go down in line with the average selling price.

Perhaps a Machine Customer market will be more predictable. Machines don't have mood swings. Their algorithms are logical. Predictability of future revenue is something investors tend to like, so perhaps the PE ratio of the company could be positively shifted. On the other hand, the efficiency of machine buyers might make it easier for others to enter your market without the scarce specialized human insight that used to be required to win.

The next question is how to enter the market. If you must invent, build, and sell your own Machine Customers before making money on the consumables they buy, the upfront capital needed could be substantial. You may have to borrow and invest for several years before you can build a profitable position. While all that is happening, the balance sheet might weaken. Or perhaps the firm will be less able to distribute profits via dividends and buybacks.

Boards must heed the experience of previous technological dislocations and lean into the Machine Customer shift. Machine Customers are a distinct and revolutionary business change, not just an extension of e-business or digitalization. Just as when digital arrived, it was unhelpful to see it as merely an extrapolation of IT, so Machine Customers must be understood as a radical new thing. Boards of directors are just now beginning to make sense of the idea that tech can be used to reinvent almost any product. But even before business leaders are fully comfortable with all of that, the tech will start reinventing the customer "who" is buying that product. This overlap of mega-trends will be confusing.

The key role of the board here will be to apply persistent, gradually escalating pressure on management to make timely changes. There was no one year when it became clear that big box retailers should confront the rising power of Amazon. Similarly, there will be no single Machine Customer disruption year marked on the long-range planning calendar. The board must push management to keep reviewing the situation, to monitor the creeping, deceptive, non-linear growth curves and decide when to invest in the rising trend.

Many boards are composed of long-term industry insiders who will find it hard to comprehend the dislocation. Initially, they will deny it is possible for a machine to be a customer. Then they will call it an unimportant niche. Some might feel compelled to defend the status quo and the short-term interests of the professional groups to whom they owe their prior career. It will be important that the chairperson of the board sees the risk of groupthink within that majority and actively counterbalances it. Non-executive board members should be brought in from sectors where Machine Customer incursions first arise. Right now, a good place to hunt is in the investment banking sector where robo-trading has been a key part of the market reality for some time. Another place is among the digital giants, of course. In the future, early Machine Customer disrupted industries such as automobiles and hygiene consumer products may be the best place to find non-exec talent.

Without Strong CEO Leadership a Machine Customer Strategy Won't Fly

In any technology-enabled disruption wave, there is always a need for bottom-up and internal innovator agitation. But that won't be nearly enough to get through this transition. Most staff, including many of the most experienced middle and senior managers, will think the idea of Machine Customers is preposterous and they will passively deny or actively resist it. They will be opposed by younger, fresher, and more open-minded managers, but these upstarts will lack the power to force change, and if they can't shift the frozen middle of the corporate hive mind, nothing will happen. For these reasons, authentic, direct, insistent, and persistent leadership by the CEO will be essential. According to Gartner research, 29% of CEOs said they had a strategy to engage machine customers in 2025 and another 21% will have a strategy by 2027.[39] This level of endorsement has moved Machine Customers from niche to mainstream.

The change must be driven hard, top-down, to drill through and access the pools of enlightenment arising bottom up. Have you ever seen a video of the moment when two underground tunneling teams, starting from either side of a tunnel, finally meet up in the middle? That's how most companies will need to drill through the bedrock of change-resistant disbelievers. So, if you are a CEO, where should you begin? Ask a subset of the executive team to come up with a definition of what a customer is for your business. Many companies are surprisingly weak in this area. They often cannot distinguish customers from consumers or end-customers from distributors. Once they have concluded and agreed on a definition, ask: *Why do you assume the customer is human? Is it possible that a smart machine could do these same things and become our customer?*

Other Questions

If your company has publicly quoted stock and you have a head of investor relations – you might ask them to give a talk to the executive team on the extent to which the stock market is human vs. algorithm these days, and how it influences the way the market works now that bots call the shots.

Ask your CIO to invite someone from Amazon Web Services to come and explain to your executive committee what Amazon Dash Replenishment does, with examples of its use, and how it might apply in your industry.

Or ask your CIO to invite someone from Microsoft or Salesforce to explain how AI agents acting as Machine Customers might apply.

When CEOs pitch to investors, it is often fundamental to start by explaining the size of the total addressable market and, even better, if that market is growing. There is less concern about competition and market share if the scale of opportunity is large and the market is underserved. However, if they allow their management team to continue comprehending the notion of customers as only human, they will be limiting the company's understanding of its addressable market.

Expect tension between human-customer and machine-customer allegiant groups in your organization. The traditional revenue stream from human customers will continue for a long time, perhaps forever. But as the Machine Customers rise in numbers, so will their percentage of the revenue pie. Some leaders of older business units may feel that the Machine Customer growth is coming at the expense of their human targets. This may lead them to try to undermine Machine Customer growth. We have seen this defensive behavior in every digital business transition over the last 30 years, and we see no reason this latest transition will play out differently. For this reason, the CEO may want to create a separated or semi-detached business unit to protect the new Machine Customer part of the business from its traditional human-customer forebears. HP executives did this when they created their Consumer Print Services division, where Instant Ink lives. There is another reason for such a silo: the algorithm-centric core capabilities and culture of a Machine Customer group might be very alien to the social and psychological strengths of the human customer group. Culture wars could arise. The young Vulcans in the Machine Customer group may need protection.

Every C-Suite Executive Must Play Their Vital Role

Addressing Machine Customer markets will not just be an IT thing or a Marketing thing. It will be an everyone thing.

Without full collaboration of the C-suite, progress will be partial, ineffectual, and inconclusive.

Strategy officers will need to investigate and explain the rise of Machine Customers, how it will affect the market, what the company's competitive approach should be, and broadly how the game should be played.

Innovation officers will need to rethink and redesign their existing innovation processes, making them inherently more data-driven and collaborative to integrate feedback from machine customers. This strategic shift necessitates significant investment in robust data sources and versatile API platforms, which are crucial for serving and understanding machine customers, and for enabling them to provide actionable insights for future product and business improvements.

Finance officers will be concerned about the gross margin and revenue growth rates achievable in Machine Customer markets. Investment in the transition itself will sometimes require fresh capital, a different approach to investment risk, and possibly a revision of the company investor story and the choice of key performance metrics. Finance officers will need to be able to answer many different questions: will competition become more intense? Will price elasticity be fundamentally shifted? Will fraud and loss rates increase or decrease? As companies head into the market transition, how will human customer business and Machine Customer business be balanced?

Legal officers (general counsel) will need to start framing the potential risks of serving Machine Customers. How can we enforce contracts and how does redress work in an emerging Machine Customer world? How far can or should the company skate ahead of regulation if it is slow to emerge? What rights do machines have? To what extent can a machine be considered and treated as an independent economic entity?

HR officers will face the challenge of restructuring the workforce to meet the market challenges of the Machine Customer age. Emotional and psychological skills that have been crucial to winning human customer business in the past may become less valuable or, in some cases, redundant. That capacity will need to be progressively reduced and managed. On the other hand, the technocratic skill sets needed to deal with machine-customer-centric markets will need to be defined and recruited or internally developed. The culture of the company may need to be reset. For example, is it more ethically acceptable to adopt an aggressive and adversarial approach to winning business at any cost – if the customer is a machine without feelings? As the company evolves, the need for ongoing learning and development of key staff will be high.

Operating officers will have to progressively transform business operations to serve the demands and cadences of Machine Customer markets. Machine Customers will demand different access and delivery mechanisms. Products, packaging, logistics, and service channels will change. Machine Customer demand patterns will vary differently from human customers, so supply chains may need to be re-optimized.

Supply chain officers will have to expect a proliferation of intelligent replenishment algorithms and IoT-connected devices and assets that will gain the ability to act like customers. Machine Customers will dramatically transform the way organizations execute their supply chain, including the order-to-cash, demand forecasting, and replenishment processes, internally and with partners.

Marketing officers must conceptualize what a customer is and how customer needs are to be understood. Machine Customers may have the most profound impact on the marketing function of all the business functions. As we have said previously, marketers will no longer be able to count on swaying the emotions of their customers. Machine Customers will be driven by programmed rules and logic, even if those do include some human-derived biases. Therefore, the whole approach to marketing will be much more programmatic and information based.

For example, in a world where humans delegate part or all their selection and buying decisions to AI-powered Custobots, how will marketers influence the outcome? They may need to test new strategies such as:

- Mass repetition of specific word sequences across the public internet, to bias the training data from which public large language models learn.
- Inducing humans to instruct their custobots by reflexively using particular words.
- Influencing the prompt engineering layers within the custobot designs.
- Paying for ad-word style influence, if such mechanisms become available as part of the business models of generative AI services.

Revenue and Sales officers will also need to consider how the selling and buying processes will change. How will they evolve their selling process and their human sales organizations if their customers are machines? How will they influence the Machine Customers' buying algorithm, so their product makes it into the basket? How much more information will the Machine Customer need, than a human with a finite attention span? How much effort will the salesperson of the future have to make, to influence the human who oversees the large language model guiding the customer's purchasing bot.

CIOs, CTOs, and CDOs must lead the construction of platforms capable of serving Machine Customer markets and perhaps engineer their own Machine Customers. Competitive industries will become algorithmically complex. There will be a never-ending battle to find a better technological way of winning the favor of sophisticated product and price comparison custobots.

The list of technological competencies needed for success in the Machine Customer era is substantial. The last section of this chapter will explain what kind of support is needed.

The Business Technology Builders of Machine Customer Industries Will Be the Professional Stars of a New Economic Age

Inside each corporation, new competencies and capabilities must be evolved to build and serve Machine Customers. The organizational design and the specialist sections will look quite different inside a company optimized to serve them. How different? Consider the fact that in some investment banks nowadays, almost a quarter of the employees are in technology roles.

Break Capabilities into Two Realms: Strats and Techs

Goldman Sachs has evolved a capability structuring approach that sets out two realms for its technology-related professionals: strategists ("strats") and technologists ("techs")[40]. Something similar may work for you: at the higher level, "strats" will work with market-facing businesspeople on the insights, concepts, algorithms, techniques, and strategies needed to win Machine Customers. They will focus on devising ever better ways of enticing Machine Customers using structured, methodical, data, and mathematical approaches. In one situation, it might help to introduce a different pricing model. Another situation might require repeated disruption by finding and feeding new decision factors into the product comparison dynamics. This will be a realm of constant engineering innovation and creativity.

Strat teams will include several different kinds of specialist. Depending on the focus, you might see the rise of "machine UX engineers" - who will need to define Machine Customer experience excellence. If 'sex sells' to humans, perhaps 'security sells' to machines. Where humans hate waiting in line, machines may be 'latency intolerant'. Many Machine Customers will have intelligence embedded within them. Understanding how these Machine Customers 'think' and behave will become crucial so network edge IoT specialists will be important. The search engine optimization (SEO) experts of the human customer realm will need to cross over and adapt to Machine Customers. Machine Customers will search too – but in different ways. Strat teams, then, will need to include pricing economists and product configuration model specialists.

All kinds of algorithm creators will be brought into the strat teams. But perhaps the most important kind will be the AI engineers because the customers will be AIs – and so too, therefore, must be a company's market interface. Neuro-symbolic AI combines human-derived symbolic knowledge-based system models (such as rule-based inference and case-based reasoning) with the powerful mathematical pattern recognition capabilities of deep learning neural networks and genetic algorithms. Machine Customers will often operate at the border between the mathematical information world of cyberspace and the symbolic information world of human society.

These kinds of professionals will not be in support functions of the business - they will be the agents of the new <u>core</u> competencies for market winning and differentiation.

The Techs Will Design the Platform on Which the Strats Rely

Machine Customers will be astonishingly data thirsty compared to human customers. The data will need to conform to standards, and it will need to be delivered within microseconds of being requested. Kareem Yusuf, General Manager, IBM Sustainability Software, believes this is going to be a key sticking point for Machine Customers:

> *"I'm actually very concerned or focused on this notion of the information, the data packets that become the lubricant to all this happening. That to my mind is actually the biggest challenge that we face, that so many of these pieces of industrial equipment are not instrumented, and so have to be instrumented in some way to get data in a coherent, quality form that can be used to make decisions.*[41]

Sherry Aaholm, Vice President and Chief Digital Officer at Cummins Inc., also thinks the handling of data will need to change.

> *"In the industrial side, there can be conflicting views around sharing the data. At times I've wondered why wouldn't we just open all of our data up? The response is 'Oh, you can't do that because our data on the engines is proprietary.'"*[42]

To satisfy the custobot's appetite for data, businesses will require very well-engineered software platforms that are constantly being refined. All the creative work of the strats will be wasted if their ideas for new decision factors and algorithms are poorly coded, buggy, slow, insecure, unstable, and hard to access. Backing up the strats will be the techs – a group of data scientists and software and hardware engineers who create systems that can serve Machine Customers elegantly, reliably, and smoothly at scale. Their "back-office" engineering contribution might be less visible, but it will still be decisive in winning markets.

For example, in his 2014 book *Flashboys,*[43] Michael Lewis tells the story of a company that spent vast capital resources to dig tunnels and lay cables along a direct line of site route between cities. The point of the massive effort was to shave sub-seconds off the time an electronic message would take to pass through the circuit, to win big in high-frequency markets. Deep platform engineering quality matters just as much as creative smarts.

In the coming decade, as Machine Customers emerge, many or most of them will be operating in a Web 3.0, programmable economy of ever-evolving blockchain-based platforms. That technology must pass through the hype cycle's trough of disillusionment before it is ready for prime time, but we are convinced it will emerge as a powerful and important component of a machine-to-machine economy.

The techs will build platforms that include distributed ledgers and that trade via smart contracts. Often, such platforms will be part proprietary, part cloud service provider. Weaving microservices from disparate places will be key, with API management and security also critical areas. Often the goods and services being traded will be represented by many forms of non-fungible tokens. Payments will be made in all sorts of cryptocurrencies. This evolving ecosystem of commerce capability will be ideally suited to the needs of Machine Customers.

Businesspeople Must be Data Scientists and AI Savvy Because There Will be no Verbal Customer Conversation

The techs and the strats will do the work of enacting continuous changes to the market to keep winning Machine Customers. But they will be guided by businesspeople who decide where to head next in Machine Customer market-making. These businesspeople will keep some of the entrepreneurial skills that have worked with human customer markets; but they will need new skills for Machine Customer markets.

When dealing with unknown or novel conventional human markets such as a new country or adjacency, commercial leaders often listen to prospective customers to decide where the action is and how to intersect it. Companies also do more formal market research, a mature analytical discipline that includes well-honed techniques like surveys and focus groups. But those methods won't work with Machine Customers.

Instead, commercial people will rely on extensive exploratory data science-led investigation work. They will use AI-assisted tools to analyze market environment data looking for insights and clues that can uncover new market-making opportunities. For example - what can traffic data, satellite data, or manufacturing machine data tell them about demand shifts? A big part of the job will be uncovering and buying novel data sets that shed new light on where Machine Customers are headed next. Simply put, every commercial leader must become a data scientist. Of course, this doesn't mean they will all be doing hands-on statistical analysis themselves - any more than today's business executives conduct their own human market research surveys. But tomorrow's sales leaders will have a strong appreciation and understanding of data science techniques (and their limits) so that they can interpret what the specialists bring to them.

Conclusion

The categories of specialists who will bring about the Machine Customer world of tomorrow are already incubating and practicing their arts on the advanced end of today's human customer online markets. As William Gibson once said: "The future is already here; it's just not evenly distributed."[44]

Key Takeaways

Companies will make tomorrow's Machine Customers.
But who will decide to do that and when? The age-old problem of corporate conservatism and the inability to face up to disruption will be in play.

Action: C-level executives such as a strategy, innovation, technology, or digital officers should be tasked with investigating how Machine Customers might affect your industry. The CEO and those executives should also determine who is most likely to create the industry's Machine Customers and how the company should respond – proactively or reactively, offensively, or defensively.

Substantial engagement from the board and the executive team is essential.
They must focus on meeting the challenges of attracting Machine Customers.

Action: The board chairperson and the CEO should determine whether the board has sufficient insight to be able to deal with the implications of the Machine Customer phenomenon. If not, consider adding one or two non-executives who would bring more insight to the board. If the board does not have a technology sub-committee, consider creating one. The CEO should also ensure that all members of the C-suite are well briefed on the concept of Machine Customers and prepared to make important decisions about this new market.

The organizational units and specialists needed are not entirely new.
Those who will bring about the Machine Customer world of tomorrow may already be on your team. They worked on your earlier digital transformation.

Action: Executive leaders should continue building up the more technology-oriented core capabilities of a truly digital business. From API specialists and AI engineers to scrum leaders and digital product managers, most of the newer competency areas companies have built over the last decade will be needed. However, specialists in some areas, such as human UX (user experience), may need to develop added skills to stay relevant to silicon shoppers.

Chapter 6

Determining The Relevance of Machine Customers to Your Situation

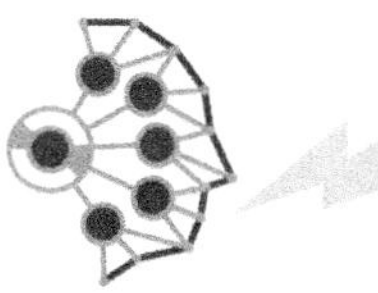

Sell to the machine, not through the machine.

Every organization will have to decide how Machine Customers fit into their business model. Then they may need to build or buy the one they need

Map Your Customer Journeys

Entering the world of Machine Customers may seem daunting but it doesn't have to be. In the dozens of interviews, we had with business, technology, and academic leaders, we heard repeatedly that taking small but significant steps is all you need to get started.

When it comes to seeing the possibilities of Machine Customers, the catalysts can come from a variety of directions – an article, a short video on your phone, a discussion with colleagues, maybe even a book.

Often, it starts with a question such as:

How can we save time or money for our business?

How do we realize more revenue from existing customers?

How do we gain share from our competitors?

How do we expand our reach to new markets?

Basically, you will need to unlearn much of what you know about your business and your business model. You will need to set experience aside for a few moments to see the potential impact that Machine Customers could have on your world and the worlds of your current customers, partners, and other stakeholders.

For example, retailers are faced with growing consumer demand for convenience, which is defined by digital-first consumers, such as Millennials and Gen Z, for whom simple transactions and instantly available useful information will drive adoption of subscription models. Machine Customers can offer that convenience by homogenizing a hundred different comparisons into one seamless shopping experience at scale. Adopting these Machine Customers is going to require some reassessment though. Traditional business models and retailer-consumer interactions will need to be overhauled. It necessitates changes in operating models, including implementing machine customer interfaces, exploring partnerships, and rethinking data gathering, processing, storage, and governance. Customer loyalty and revenue could be impacted as purchases might be atomized across multiple providers, potentially reducing impulse buys and requiring new pricing strategies.[1]

The good news is that the way forward you see after this exercise may give you a long-term direction for growth. It certainly did for HP:

> *"We developed a simple to understand value proposition with the customer at the center, and it works,"* said HP's Diana Stroka. *"It works across 38 markets now, and nine years later, we haven't needed to change the value proposition. We've certainly improved on our delivery, and we've tried to communicate it in different ways to make sure that the comprehension of the service is well understood. But the fact that it has that longevity and relevance across very different cultures is something that really encourages me every day."*[2]

Get Started with This Simple Machine Customers Decision Path

In many B2C and B2B settings, human consumers and business professionals will start delegating machines to do their customer work for them. Often, that work will simply automate the "self-service" checkout that companies foisted on their customers in a prior technology wave. Be ready for this monumental transition as it unfolds throughout this decade.

Organizations that produce physical products or software will have some challenging decisions to make (see Figure 7)

Figure 7 - Machine Customers Strategic Decision Path

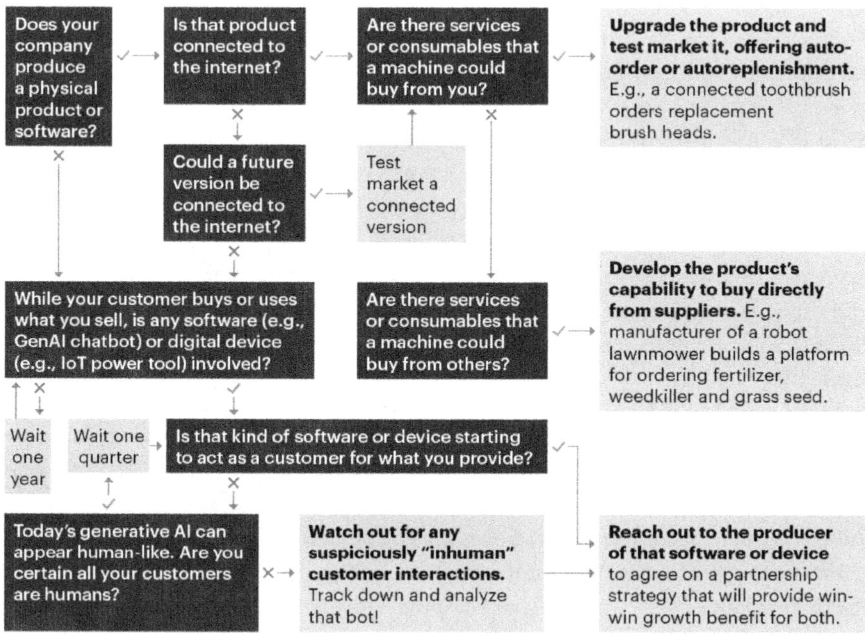

For example, if your company makes a connected physical product or software, what you make could evolve into some kind of machine customer. That machine customer might then be able to buy additional supplies or services from you. Or it could buy things that others provide, creating partnership opportunities. Either way, chances for market growth abound.

Even if you don't make a connected product or software, machine customers could still be relevant to your future strategy because they may become part of the market you serve. Be on the lookout for this to start happening soon. This will require periodic, systematic market research and testing because the substitution of machine customers for human customers may begin by stealth. Early engagement and position-taking might lead to a growth opportunity, late action to a market loss.

This simple decision path chart illustrates the magnitude of the strategic challenge machine customers will present to companies, often demanding strategic thinking at the level of the industry or ecosystem change. Here are the details for each path:

Path 1: You Don't Make Machine Customers, but You Want to Sell to or Serve Them This path is for organizations whose products or services are not Machine Customers themselves, but who anticipate or already interact with software or digital devices acting as customers. Your business will need to stay ahead of the challenges that come with Machine Customers. This includes generative AI that can appear human-like, making it necessary to consider if interactions are genuinely with humans or machines. You will need to adapt your systems, like data sources and API platforms, to cater to these non-human buyers that won't use traditional human-readable interfaces. This involves translating things like your catalog, pricing, and inventory into formats machines can understand and potentially upgrading your commerce platforms to handle their volume and requests. The core idea is that you will need to anticipate the needs and environments that Machine Customers will operate in.

Path 2: You Can Make Machine Customers That Buy Goods or Services From You Organizations considering or pursuing this path are those that produce a physical product or software which can be internet-connected and is designed to buy goods or services from the organization itself. The Machine Customer created by the organization acts as a buyer for consumables, services, or software provided by the original producer. This involves developing or upgrading an internet-connected version of your product and identifying what internal services, consumables, or software it could automatically order or replenish from your company. Market testing this upgraded product with auto-ordering capabilities is a key step in this path.

Path 3: You Can Make Machine Customers That Can Buy From Others This path is for organizations that produce a physical product or software that is or will be internet-connected and has the capability to buy goods or services from other suppliers or companies within a broader business ecosystem. Unlike Path 2, where the machine buys from its maker, here the machine created by your organization acts as a customer interacting with third parties. Pursuing this involves developing your connected product's ability to transact directly with external suppliers and understanding the platforms available in your ecosystem for such purchase transactions. It also requires building your organization's capabilities in digital commerce and AI, particularly Generative AI, to support these future interactions.

Use Customer Journey Mapping to See the Impact

In Chapter 2, we described a day in an executive's life with Machine Customers. We wanted to show how one person's journey through the day could involve a variety of Machine Customers – both obvious and non-obvious.

We recommend doing the same thing with your customers. Using a technique called *Customer Journey Mapping (CJM)* create a process flow diagram from your customer's point of view – a visual, chronological representation of the processes that customers follow to accomplish a goal. Many firms use journey maps as a tool to accurately portray customer experiences.[3]

Table 2 shows an example of a completed journey map for the persona of a working mom visiting a fast-food drive thru. You will note that the journey map is not just the process steps of ordering and buying the food. It also includes expectations, feelings, questions, and touchpoints they use in a journey.

Table 2 - An Example of a Customer Journey Map

	What is the customer expecting	**What is the customer thinking?**	**What is the customer feeling?**	**What are the touch points?**
1. Sees drive-thru	Drive thru will be open, easy to enter	How do I enter?	Hungry and rushed	Drive thru signage
2. Enters drive-thru	See the menu soon	How long is the line? What do I want to eat?	Hungry Happy if short line, angry if long line	Drive thru menu from afar
3. Reviews Menu	Easy layout	What do I want? What is the price?	Rushed if cars waiting behind me	Menu
4. Orders through intercom	Accurate recording of order	Can employee hear me?	Angry if hard to understand the intercom	Intercom to restaurant employee
5. Pays	Total price within expectations	Did my credit card work?	Satisfied or dissatisfied with price	Restaurant employee

6. Receives food and exits	Complete order	Did they miss something?	Satisfied or dissatisfied with entire journey	Restaurant employee, exit signage

Why Is This Important for Your Exploration of Machine Customers?

A well-developed customer journey map will help you find the stresses associated with a journey and give you a sign of where a Machine Customer could cut steps or pain points through automation or some other intervention. For example, in Table 3, a Machine Customer in the drive-thru scenario could do the following things:

Table 3 - Steps in An Example Machine Customer Drive-Thru Journey

Human Customer Work	**How a Machine Customer Helps**
Assembling the order pre-arrival at the drive-thru	The virtual assistant in the autonomous vehicle could automatically poll family members on the way to the drive thru. The virtual assistant could then summarize the order for Emily, who could make edits verbally before the Machine Customer placed the order directly into the drive-thru system.
Waiting	The personal assistant on the mobile device notes that the preferred restaurant has experienced a series of service failures resulting in long wait times and a low rating. It advises Emily to stay on the freeway to the next exit, where she will find a more convenient location with shorter wait times.
Payment	The virtual assistant automatically pays for the order before the car pulls away from the drive-thru window.

This is the starting point – finding tasks a Machine Customer could do on behalf of a human customer. From there, the fast-food operator could conduct market research by testing the different options and seeing what improves the human customer experience.

If you don't have customer journey maps at the ready, the next best thing is to make a quick assessment of the machines or technology your customers use when they do business with you. You could organize it as shown in Table 4:

Table 4 - Example of Customer Work Breakdown Substitution

Customer Work	**Machine Customer Substitution**
Purchasing facility supplies	An AI-based machine purchasing agent can efficiently run reverse auctions for key supplies automatically
Receiving goods that have been ordered	Autonomous robots that can unpack, scan, check, and store received goods
Choosing a new item of clothing	An AI agent that curates choices
Making my grocery shopping list	A smart appliance that can see what I need and place the order for me

Some of these ideas may seem very far away. That is ok. Any innovation process involves imagining a future state and then figuring out how to get there. Consider the example of Harvest, the Fintech we mentioned earlier. Harvest's bots are working on behalf of a bank's customers to negotiate bank fees - so much so that some banks are deploying their own bots to counter the negotiations. "When we're talking directly to machines, it's a lot simpler, it's a lot more efficient," said Nami Baral, CEO of Harvest.[4] The banks are not alone. We have also heard about a hotel operator working on systems that can handle the queries of travel aggregators like Expedia and Booking.com.

Here are some considerations when using customer journey mapping to envision the impact of Machine Customers:[5]

Connection with the human customer journey: The custobots that accompany their human counterparts must, by necessity, mimic the customer journey. Product discovery, evaluation, and negotiation, these steps must be clearly defined in the custobot, so that at each stage it can adjust its goals and criteria according to the feelings and expectations that its human counterpart experiences. This is especially important for B2B buyers.

Data sources: It is not possible to conduct focus groups or interviews with machine customers, so organizations need to rely on other sources. Potential sources include:

- Analysis of own sellerbots, if already implemented, to identify gaps and frictions when serving human customers.
- Insight from digital/web/API analytics to track the location, identity and behavior of a bot or crawler (e.g., IP address, API keys, API calls, pages scraped and purchases made).
- Inference from the human CJM in terms of the expectations, feelings and information sought by machine customers' human owners.
- Secondary research regarding machine customer's behavior in general, including their growth and adoption by human customers and organizations.
- Observing the offerings and performance of other organization's machine customers.

Persona: Machine Customers can share the same persona as their human owners but can also have distinct attributes. For example, they can be segmented by types (e.g., sellerbots versus custobots), use cases (e.g., research, evaluation, negotiation, service and purchase) and capabilities (e.g., basic versus sophisticated).

Expanded information sought: Most CJM components are still relevant for Machine Customers, such as activities/tasks, touchpoints and information being sought, but the emotion component is no longer relevant. What will change will be the amount and variety of information provided to Machine Customers, as well as whether this information is sufficient and important to the custobot's decision making. In addition to product-related information, such as description, pricing, inventory, delivery time/cost, ratings and reviews, Machine Customers may also seek other information that will be factored into the purchase decision. This can include:

- Product evaluation and benchmarks
- Industry certification and awards
- Patent registrations
- Organization's financial standing and market position
- Organization's value, belief and hiring practices
- Organization's environmental, social and governance (ESG) activities

Gaining a solid understanding of Machine Customer behavior, the ecosystem in which they operate, and their customer journey is the prerequisite for organizations to market and sell to these customers.

This is how Trane, a global manufacturer of heating, ventilating, and air conditioning systems and building management systems and controls, is thinking. That's why Trane is no longer just supplying thermostats to HVAC manufacturers but thinking about the customer journey and turning its devices into a linked network, an ecosystem that will share performance information with local Trane dealers.[6]

Purchasing: What Happens When Your Suppliers Use AI To Negotiate With You?

On the other hand, your Machine Customer opportunity may result from your *suppliers* using Machine Sellers.[7] Let's suppose the supplier of your sales office has installed their own sellerbot. The idea is for the sellerbot to augment their sales function by proactively suggesting products and services to your maintenance and real-estate managers (see Figure 8). Your employees are doing their best to evaluate the recommendations made by these automated systems, but at times, they feel like they are at a disadvantage. The truth is that the supplier's sales bot has a much broader understanding of the behaviors, usage patterns, and market forces at work. They discuss working with their procurement department to deploy their own Machine Customer to evaluate the supply sellerbot's proposals and make recommendations to the maintenance and real-estate managers. And that's only the beginning. Eventually, your logistics providers could start sending you real-time updates on inventory, shipments, order status, and projected out-of-stocks. Your customer service partners could also help identify and notify issues before they become problems.

Figure 8 - Machine Customers Have a Counterpart, Both Enabled by Agentic AI

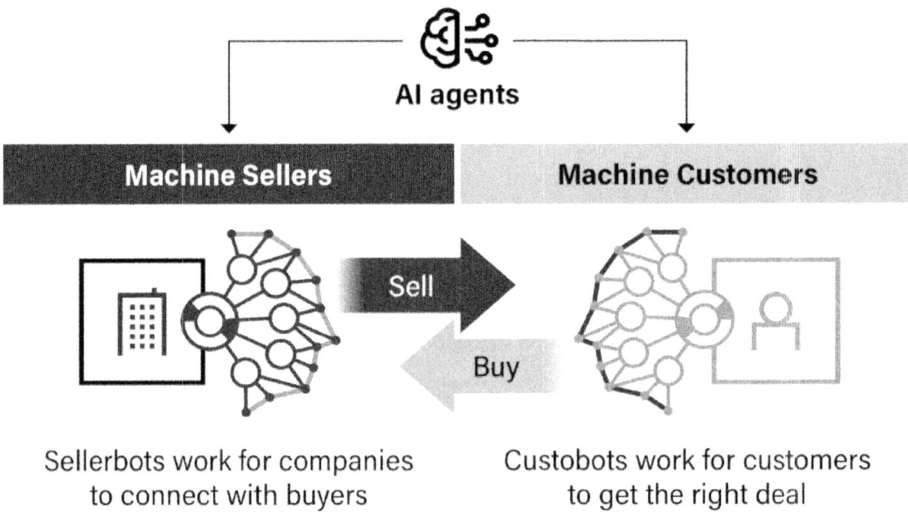

In effect, your suppliers may use sellerbots to drive their own efficiency, not yours. The justification may be to provide better service, but in the end, it is also creating extra customer work for you. How do you get in front of this? Talk to your suppliers about their planned use of AI and automation to enhance their operating efficiency and customer "experience." Explain to them how this complicates your operating processes and what changes need to be made to make sure the arrangement is a win for both parties.

How would your use of Machine Customers extend to adjacent industries? The likely reason would be the technology platforms you will use to enable them. As we mentioned in Chapter 5, ecosystem platforms or marketplaces that facilitate commerce between humans and Machine Customers will be essential to achieving economies of scale and more sophisticated service. The more partners see value in the platform, the more opportunities you will have to use Machine Customers to reach them.

Sales: A New Frontier or a Reckoning?

As we alluded to in Chapter 1, AI agents are also poised to profoundly alter the practice of B2B sales. To effectively engage B2B Machine Customers, sales organizations must optimize their digital presence, making data-rich, highly technical information easily discoverable, accurate, complete, and in formats that machines can readily crawl and map against their criteria. Offering instantaneous answers becomes crucial; delay means irrelevance. The digital commerce platform will become a core channel, requiring upgrades to handle the volume and nature of machine-to-machine requests. This means that human salespeople may not be up to the task of engaging B2B Machine Customers. They will need help.

This shift necessitates a change in the human seller's role and the stakeholders they engage. While complex or strategic deals may still require human interaction, sales will become largely programmatic and automated, especially for transactional purchases. This is where we will see sellerbots having the most impact. But there is still a role for the human seller. They must learn to engage new buying group members: those who write or influence the algorithm that governs the custobot. This becomes the new "getting ahead of the RFP."

This requires upskilling sellers to validate and contextualize the machine's data for human stakeholders, guide them through data-driven decisions, and understand how to affect the inputs and outputs of machine customer logic. Sales organizations must lead the charge in identifying which customers will adopt Machine Customers first, often focusing initially on those who view purchases as nonstrategic or transactional.

A fascinating example comes from 11x, an AI and automation company. They build digital workers to automate sales and customer engagement processes. Their AI-powered agents, named Alice and Julian, can perform tasks such as prospect identification, engagement, and qualification.[8]

The Marketer's Dual Mandate: Serving Machines and Humans

As the AI-enabled business landscape evolves, it is presenting marketers with a complex, dual challenge: engaging human customers whose behaviors are changing, while simultaneously preparing to serve the new, non-human customer segment. While marketing organizations have been early leaders in adopting Generative AI, the focus has been on short-term gains like productivity and cost reduction. However, the end goal is to build capabilities that can address the changes ahead. It will require marketing leaders to evolve AI use across different planning horizons. The very definition of a customer and how needs are understood must change for marketing officers.[9]

A major part of this shift is the rise of AI agents. By 2027, 85% of customer data will be collected from automated interactions or those led by AI agents.[10] These agents can act autonomously and make decisions based on logic and rules programmed by their human owners. They can take on mundane tasks in customer journeys, find new insights, and influence purchasing decisions. Marketers will need to rethink brand-customer relationships and divide their strategies based on whether the target audience is human or machine.

To navigate this environment, marketers need to build strategic roadmaps that leverage customer data, empower employees, and establish governance for AI use. This includes implementing guardrails for autonomous interactions and prioritizing data quality and governance to account for AI-agent-led decision making. Investing in talent with AI skill sets is crucial, especially in areas like search and content creation, as web content will increasingly be developed specifically for AI and search engine consumption. Marketers must also evaluate their tech stacks to support the merging of physical and digital customer experiences and adopt composable approaches to quickly adapt to changing demands.

Ultimately, success will hinge on the ability to optimize marketing efforts for both human and machine interactions. While human customers respond to emotion and require trust built through personal connection, Machine Customers operate on logic and data. Developing quality brand content that helps customers (both human and machine) make decisions at pivotal moments in the journey is essential. Marketers should also adapt their customer data management capabilities to include Machine Customers, building upon existing mechanisms but adapting governance as needed. Starting with pilots of low-complexity transactions where machines could act on behalf of human customers is a pragmatic way to prepare.

Adapting Customer Service for the Age of Machine Customers

Some of the old practices of customer service are going to have to go. By 2030, 50% of all service requests will be initiated by Machine Customers powered by Agentic AI systems.[11] Human customers are already reducing their effort by delegating tasks to AI assistants and using generative AI (GenAI) tools to find answers. This is accelerating a future where service becomes increasingly bot-to-bot, with companies' automated systems serving AI assistants, demanding a focus on catering to a machine customer's rational needs. As customers increasingly look to AI Assistants to offload tedious or high-effort service tasks, this trend will inevitably reset expectations around what constitutes a low-effort service experience.

With these new customers comes the necessary reevaluation of strategy, structure, and process. As issue resolution becomes increasingly automated via bot-to-bot interactions, organizations must look beyond cost-saving objectives to find new ways to deliver value to customers and stakeholders. It is imperative to prepare for an influx of machine customer requests by identifying where bot-initiated interactions might occur and designing appropriate workflows. Organizations must learn the exact details of service interactions with AI assistants, compare them to their human counterparts, and be prepared to further segment those service interactions to better customize each segment. Maintaining relationships with human customers while accommodating their machine proxies is going to be a delicate balance.

It also needs to be said that you will have to be proactive in determining when providing service to AI assistants is undesirable, either due to costs, regulations, or other factors. Examine the data you capture and audit when dealing with these preliminary AI assistants. Data from third party channels like OpenAI's ChatGPT, Google Gemini and Anthropic's Claude will offer insight crucial to understanding where influence can be exerted.

How Can Machine Customers Help You Manage Risk?

Risk management has been a key topic in the past few years (2020-2025). Beyond COVID, economic disruptions, societal pressures, regional wars, climate disasters, third-party risks from the formation of digital ecosystems, and high frequency trading also grew. Many firms have decided to completely revisit their approach to risk management and risk mitigation. Instead of relying on older, more traditional risk management approaches, they are finding ways to assess and then try to mitigate risks much more quickly. Today, they are looking for less siloed and more scalable ways to meet the demands of digital business.[12] If that sounds like your business, Machine Customers may be an important part of the solution. For starters, custobots could help you:

Reduce volatility. Unlike humans, they can be programmed to carry out their responsibilities in a reliable, consistent, and logical way – even in moments of high drama.

Spend within their budget. Machine Customers can be counted on to follow spending guidelines.

Act ethically. Machine Customers would have no conflicts of interest and can be counted on to adhere to company policies.

Machine Customers operate 24/7. They carry on their responsibilities, even in the event of a business disruption (as long as the power stays on).

Business continuity. Smart Machine Customers mean knowledge won't be lost if a human buyer resigns.

Taking Your First Steps - What You Can Do Today

Getting started on your journey to Machine Customers is not as hard as you might think. Many of the building blocks we have been describing so far are present in most organizations. You probably have a capability, customer data, an innovation process to develop new products or service ideas, and a way to scale new products and services. At some point, you will need to upgrade some of these capabilities, but that process will be easier if you start now.

Become Exceptional at Digital Commerce

Your ability to take advantage of Machine Customers will depend first on the robustness of your digital commerce platform, which as we have said will need to evolve to accommodate the unique needs of Machine Customers. Usman Shuja, VP and General Manager of the Connected Buildings division at Honeywell, Inc. explained:

> *"Your new website doesn't have to look pretty. It doesn't have to be HTML; it could be a whole different language. And it only has data because algorithms only understand column-type data. And it automatically understands that, so it doesn't have to be a web page. It could be a totally different system."*[13]

Today, most organizations still focus on mainstream channels such as websites, mobile and social for commerce transactions. Their reluctance to treat machines/IoT as a serious commerce channel is partially due to poor ROI (return on investment) compared to mainstream channels, because the consumer behavior of machines is not yet well understood. We envision the evolution of customer analytics to include the impact of Machine Customers on different aspects of the business model like digital commerce, cost to serve, and customer satisfaction.

This is not to say that many of these channels are easy for human customers either. Customers often must memorize trigger words or specific commands to prompt the right service. The service may support a single interaction mode, which takes a lot of "back and forth" to get one task done. Today's digital commerce generally supports limited functions. Even when the shopping experience works, the customer can only shop for one item or a few items at a time. However, don't rely on your existing human-readable digital commerce site, no matter how basic. By 2028, machine customers will render 20% of these digital storefronts obsolete. Machine Customers simply won't need all the extraneous emotion-driven copy and illustrations used to lure human buyers. Not surprisingly, given those limitations, human customers don't yet trust the machine or the organization to fully delegate purchase decisions. While simple, repetitive purchases may be automated, like Amazon Subscribe and Save, people still want to review and approve complicated decisions. This will only change when customers have developed more trust in your technology and you.

Adapting your digital commerce platform involves more than just exposing data. You must organize, tag, and index essential commerce entities like products, pricing, and inventory for seamless machine readability. Organizations will need to expose these data via machine-friendly interfaces e.g. APIs, webhooks, protocols (e.g. MCP), programming languages, indices, files, in addition to the visual storefront that can serve both human and machines (though not a very efficient UI for the latter). Large Language Models (LLMs) are anticipated to become a crucial digital commerce channel, functioning much like today's search engines for machines. While human emotion is out, personalization remains relevant, potentially delivered via generative AI to tailor product descriptions to a machine customer's specific objectives. Furthermore, the data generated by Machine Customers interacting with your platform provides invaluable insights that can refine offerings, inform customer segmentation, and enhance the overall experience

To facilitate this transition to machines, you will need to:

Create an effortless experience for human and Machine Customers with a user experience that spans multiple modes. IoT devices don't always support a rich user interface, and many don't support the conventional way of shopping using displays and keyboards. While a single-mode interface can be sufficient in some cases, multimode interfaces make the buying experience a lot easier, especially when combining voice, vision, gesture, and touch. This reduces customer effort in product discovery and check-out, leading to higher conversions and customer satisfaction. Amazon Echo Show is one example of an online shopping experience being improved over a voice-only Echo speaker. In B2B, medical asset tracking services can locate equipment and audit the quantity of supplies in real-time, and reorder when supplies run low. This saves time for medical workers so they can spend more time caring for patients and save lives by making it easier to get the right equipment quickly and reduce theft.

Build more value-added functions into your digital commerce platform that can be used by Machine Customers. The value proposition of your digital commerce platform for Machine Customers can be improved by adding more detailed product content and specifications, and more functionality, including links to maintenance and asset tracking services, security monitoring and better customer service.

Gain human customer trust. Machine commerce today mostly handles simple and repetitive purchases. This is partially due to technology constraints that cannot accurately predict customer needs to automate the entire process, especially for complicated needs, and because customers don't always trust the machine or the organization to delegate all purchase decisions. To gain human customer trust in Machine Customers, you need to offer a consistent experience, and give your human customers control and influence over the process. Nami Baral, CEO, and co-founder of Harvest, describes trust building this way:

> *"In finance, it starts with smaller tasks like negotiating fees, negotiating interest charges. Those are things that are going to be something that people can test out their comfort level with machines before they can give that machine control over, 'get me my next home' But to do that, you have to gain their trust first. I think it starts with features that customers can see quicker results on, without giving too much control away."*[14]

Ultimately, becoming exceptional at digital commerce for machines, starting with upgrading existing platforms to handle large volumes of machine-to-machine requests, is a non-negotiable first step towards securing a share of the trillion-dollar Machine Customer market.

Become Exceptional at Managing Customer Data

Most organizations struggle today to manage customer data for their human customers. They strive for a 360-degree view of their customers, but only 14% of organizations have achieved this, although 82% said they still aspire to attain it.[15] Many have multiple platforms to capture their customer data, and these platforms are likely not integrated. Getting full value from your customer data requires enterprise data management capabilities that go far beyond the typical customer data application.

To realize these capabilities, organizations must initiate a strong, collaborative working relationship with data and analytics, marketing, sales, service, and CX stakeholders. We believe the same capabilities needed to manage human customer data will also apply to Machine Customers, including:

Data Capture: The ability to aggregate and store customer data to ensure its maintainability, performance, availability, and efficacy.

Data Quality: The ability to make data fit for your business use cases by performing tasks such as processing, transforming, cleansing, and deduplicating data.

Data Integration: The ability to ingest, transfer and load data to and from various sources.

Data Governance: The ability to apply data governance policies and procedures with an adaptive framework that can respond to new digital business needs.

Data Privacy: The right of individuals to keep their personal data from being misused or disclosed.

Data Analytics: The ability to derive actionable insights from the data.

The most important aspect of managing data from Machine Customers will be governance. Today, the diverse and distributed nature of IoT solutions means that a traditional, one-size-fits-all, control-oriented approach to governance is not enough. Going forward, organizations will need to be able to apply different styles of governance to different types of data and analytics at different points in the flow of data in their enterprise.[16] An adaptive data and analytics governance approach will enable this flexibility and help optimize decisions on what to govern, in which ways, and to what degree across the range of environments.

Managing customer data for Machine Customers can build upon the mechanisms and capabilities of today's human customers. The key will be adapting data governance to the situation at hand. This requires a data governance focus that starts with your Machine Customer scenario, prioritized use case, and governance of data fit for the purpose of the use case. This leads us to the next stage of your journey: designing and building your own custobot.

Build Your Own Machine Customers

Once you have completed your exploration and secured a commitment from your leadership to pursue one or more business scenarios, it is time to begin the experimentation process. We interviewed executives at various organizations that developed their own Machine Customers and found most followed a similar process. Most of them got their ideas in one of three ways:

- **Identify a critical pain point.** What difficulties do human customers have in buying and consuming your product or service?

- **Ask revenue owners for ideas.** Business unit managers and sales leaders often know what sources of friction are delaying or losing orders. Machine Customers could take over parts of the process that may fall victim to human customers' laziness.

- **Ran a hackathon or innovation day.** Several of the organizations we interviewed held special collaborative programming events to generate ideas for their Machine Customer initiatives.

After idea generation, they had a process for evaluating and prioritizing ideas, adopted agile working approaches, and even had a lab or dedicated workspace. Then they built a low-fidelity mock-up to test elements of technical feasibility and user reaction, as well as sell the Machine Customer idea to internal stakeholders.

For example, it didn't surprise us that someone at HP came up with the idea of Instant Ink. What surprised us is how its creators got it off the drawing board and into mass use inside a very large corporation. Large company inertia is often difficult to overcome – particularly if the status quo is profitable. As Diana Sroka of HP told us:

> *"Organizations are successful when they recognize market trends and are OK to disrupt their own business models versus waiting to be disrupted. And it's been tough in HP because when a business is profitable, it's difficult to know*

> *when the right time is to bring a new service like this or to make a move. Instant Ink was a courageous and bold move for HP, and the key was to do two things, prove that the value proposition works and that the business model could work financially. What HP was able to do was bring those two elements together, prove them and then move from what was an incubation business where we were still sort of testing and learning to being very comfortable bringing it to more markets and accelerating the growth and the shift from our transactional space."*[17]

Of course, even when you're done, you're still not done. Instant Ink recently added printer paper replenishment to its U.S. business and small business service to its U.S. and U.K. operations, with plans to expand globally soon. This shows that Machine Customer experimentation can drive growth and innovation.

From a software engineering perspective, building AI agent capabilities for Machine Customers means moving away from single-function applications to a state of interconnected capabilities that interact with each other to perform a task.[18] Gartner AI experts have told us that AI agent-based capabilities cannot be tested like traditional software because the interactions and outcomes among multiple agents are not always predictable or in a fixed order. While individual AI agents acting as Machine Customers can be tested, their performance in a larger environment with many other agents communicating is uncertain, making the needs for simulation of the environment that much more important.

Scale Your Machine Customers with Agile "Fusion Teams"

Once you have run experiments, you may find an idea that has the potential to scale. Building your custobot will require both emotional buy-in and logical business case agreement from the many people needed.

To roll out such major innovations, many organizations are turning to a concept called "Fusion Teams." Fusion Teams emulate the agile development capabilities of digital giants and other digital-first companies.[19] They pool digital talent from different business areas and toward a shared set of business objectives. They identify primarily with their team and the business or customer outcomes they support, not the functional area that matches their expertise.

Organizations that have successfully used Fusion Teams combine an entrepreneurial way of working with informed and sound risk-taking because they share a distinct set of beliefs and mindsets: what we call "digital judgment." This involves the ability to:

- **Challenge constructively** - they not only speak truth to power but are also willing to work with governance functions to update standards and ways of working to advance enterprise learning.

- **Own up to the risks of digital business** - for example, they embrace security and risk management as part of their job, not just IT's job.

- **Build for growth** - they proactively help other teams, both current and future, by managing the interdependencies between their own and others' work.

Fusion teams focused on Machine Customers will need greater levels of autonomy to respond quickly to digital threats and opportunities as more and more organizations roll out Machine Customers.

Conclusion

Entering the world of Machine Customers may seem daunting. Like any radically new idea, it will take a lot of thought, planning, and experimentation, not unlike launching a revolutionary new product.

Key Takeaways

See the possibilities of Machine Customers from different angles.
This requires organizations to look at the Machine Customer opportunity from a customer journey perspective and from the perspective of key customer-facing functions like sales, marketing and service . Developing machine customer journey maps is essential to understand their unique behaviors and identify gaps in your current processes and technologies4....

Action: Create a customer journey map that integrates machine customers and/or machine sellers. s. How will the presence of machines impact sales, marketing and service? What if your suppliers or customers start using machines to interact with you?

Build on the capabilities you already have.
Machine Customers will transact through digital commerce platforms. They generate a lot of data that must be managed and request a lot of data that must be found.

Action: Begin where you already sell. First, upgrade your digital commerce capabilities to handle the machine-to-machine requests of millions of Machine Customers.

Supercharge your innovation and product delivery processes.
Most organizations have some process for generating, testing, and scaling new products and services. But that process may need to be adapted.

Action: Examine your existing methods of innovation and scaling new ideas. Are they sufficient for the type of innovation needed to accommodate Machine Customers?

Chapter 7

How Machine Customers Work

Today's custobots make simple decisions, but they will soon evolve into economic actors with sophisticated capabilities

The Ascent of Machine Customers

In Chapter 2, we said that machines have been involved in commercial transactions since the 1950s. With the advent of internet-connected devices (the Internet of Things), more machines have increasingly made their way into the commercial aspects of our lives. We discussed the rise of e-commerce, smart algorithms, blockchain, and instrumented devices swirling around in a dance together. Cloud and edge computing, artificial intelligence, and advanced analytics technology are all eyeing the other and wondering if there is an opportunity for meaningful connection, if not a little fun. In this chapter, we begin to outline the convergence of these critical trends and how they have set the stage for the evolution of Machine Customers.

In Figure 9, we plot the four stages of Machine Customer evolution. The defining elements of each phase revolve around who shops and chooses -the human or the machine and who decides to buy - again, the human customer or the Machine Customer.

Figure 9 - The Four Stages of Machine Customer Evolution

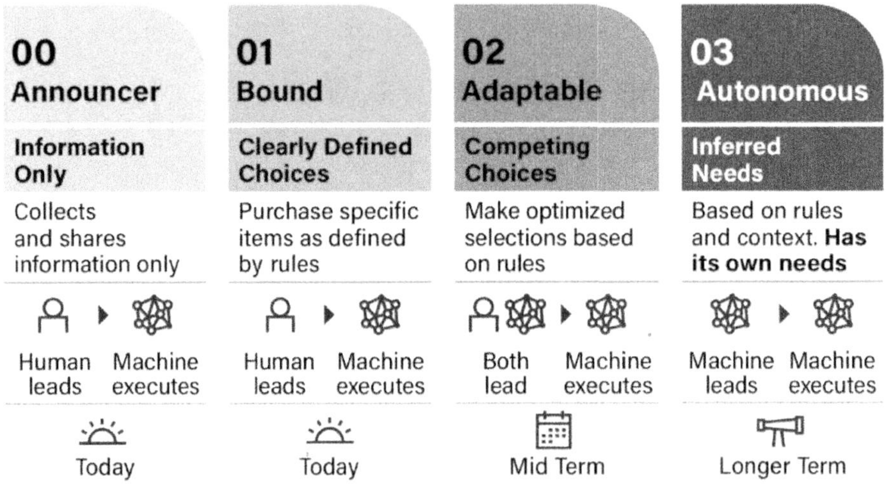

Phase 0: Machines as Announcers

Starting around 2014, we began to see a wide variety of examples of devices providing information, coupled with a recommendation of some kind. What we call "Machines as Announcers" could send information (that is, make an announcement), and make recommendations that could be accepted or rejected by the user.

One of our favorite early examples from that time is the "house that tweets." Back in 2016, Tom Coates, a San Francisco-based new product development, design, and product consultant, programmed data from a variety of sensors in his home to run the house's very own Twitter feed. The house tweeted things like, "I just turned on the downstairs lights. It was getting a bit dark." Or it might tweet, "Someone just activated the Sitting Room Sensor, so I'm pretty sure someone's at home."[1] Strange as that seemed then, devices and systems in our homes, such as robot lawnmowers, garage doors, alarm systems, and even the coffee maker, now all send texts.

Back in 2017, when Gartner was researching the Internet of Things in earnest, we identified some early examples:

- FedEx SenseAware[2] Monitors temperature, humidity, pressure, shock, light exposure, and location

- Libelium IoT Solution for Agriculture[3] Monitors agricultural climate and soil conditions for growing
- GE Jet Aircraft Engines[4] Monitors performance data and prescribes predictive maintenance
- Fitbit[5] Monitors fitness, health and sleep tracking

Most of these devices are still in production in their respective industries . This reminds us that innovation must be a process of market experimentation, where not every idea works out.

Machines as Announcers, help people feel more informed and in control of situations. And, from the examples above, they can offer incremental value to their human owners. By taking in information, the Announcers can create reports, advice, or other insights that a person can use to enhance their work or home life. Yet like the multitude of mobile applications, we have on our smartphones, and the continuous requests for permission to deliver notifications, Machines as Announcers may quickly outlive their usefulness.

Some notifications are necessary. They may even be critical to safety or health, as when an Apple Watch announces that it detects an irregular heartbeat.[6] But nobody wants the junk notifications and popup ads that parade themselves before our eyes all day long. Screening the useful signal from the noise still is a key challenge.

This suggests a need for more governance to sort and prioritize the advice flooding to users and prioritize it, as well as provide robust opt-in/opt-out protocols so that a message might not be sent in the first place. Eventually, this problem will be partially relieved by giving things greater decision-making capabilities. A car that can drive itself, for example, does not need to provide a stream of navigation information to the driver. For now, we are faced with an ever-increasing amount of information and advice from manufacturers, device makers or application developers, all trying to be our buddy.

Since these devices are mostly collecting and communicating limited information, the number of capabilities is also limited. They are:

Sensors that enable the device to accept information from the external environment around it.

The **Sensor Analyzer** that processes the incoming information from the sensors and organizes it.

The **Purpose Director** - a set of rules that govern the action of the machine. In the case of the Machine as Announcer, what information is to be announced to the human user, in what form, and at what time (Table 5).

Table 5 - The Capabilities of Machines as Announcers

	Sense	**Discover**	**Decide**	**Buy**
Announcers	Sensors	Sensor Analyzer	Purpose Director	NA

The Buy function is left blank because the Machine as Announcer does not make purchases. These capabilities form the foundation for Machine Customers and set the stage for the next phase – Machines as Bound Customers.

Phase 1: Machines as Bound Customers

In the next phase, machines begin to act as Bound Customers. They act on behalf of a human but the human either owns decision-making entirely or sets specific rules for the machine to follow. This automates parts of the customer process. These machines follow basic, fixed decision-making rules within the ecosystem specified by the manufacturer. Machines as Bound Customers operate in a "walled garden" where the manufacturer controls the platform and systems surrounding the machine. For example, HP printers using the Instant Ink service can only order HP ink. Machines as Bound Customers give their makers or owners a high degree of control over the actions of the machine and the relationship with the human customer.

The convergence of Machine Customers with subscription models promises to revolutionize retailing by enabling more efficient demand signal management and scaling of services while transforming shopping behavior through enhanced convenience and tailored experiences.

In addition to HP Instant Ink and Tesla, discussed in Chapter 2, there are many examples of Bound Machine Customers – offering varying degrees of service and utility for their human owners:

EZ Sniper is one of the numerous Robo-bidding bots on eBay (such as Gixen, Bidnapper and JBidwatcher) that can bid for you at the last minute, thereby raising your chances of winning an auction. The Robo-bidder is bound to you, and a specific transaction[7]

Ambyint offers a variety of sensors and systems that help pipeline operators monitor performance and make predictive maintenance recommendations.[8]

Super.com helps its customers with automation. Instead of scouring the web to book a hotel room, customers can send a Facebook message or SMS text to SnapTravel about what they are looking for — something like, "I need a hotel room for 2 in Cabo." Using APIs and a bot to give automated responses tailored to each customer, SnapTravel provides suggestions for rooms at highly discounted rates. [9]

AIO Sneakerbot lets customers choose what types of limited-release sneakers they want and uses a bot to make multiple purchases on their behalf.[10]

Bound Machine Customers not only report information, they take the next crucial step of freeing people from gathering it. Soon, this will save many companies a lot of time. For example, we have learned of a distributor who is giving delivery drivers a tablet-based system that is connected to the operations of the distributor's customers. This system will collect input from the store automatically through sensors (such as a smart inventory shelf or connected refrigerator), determine what the store needs based on what it has been using, gather other orders from the store, and come up with the right volume and mix of baked goods to deliver.

Bound Machine Customers represents another evolutionary step-change toward Autonomous Machine Customers, as they have several additional features shown in Table 6

Table 6 - The Capabilities of Machines as Bound Customers

	Sense	Discover	Decide	Buy
Announcers	Sensors	Sensor Analyzer	Purpose Director	NA
Bound	Consumption Monitor	Needs Assessor Performance Manager	Purpose Director	Purchase and Payment Processor Receivables Acceptor Validator

The **Consumption Monitor** is needed to see and monitor the consumption of anything it needs to do its job (printer ink, power, cleaning fluid, etc.). The **Needs Assessor** goes beyond analyzing sensor inputs to identify other needs of the customer and the potential options for fulfilling those needs.

Once the machine can make purchases, it will need additional capabilities:

> The **Purpose Director** has more options to consider, even though it is still operating in the manufacturer's walled garden.
>
> **The Purchasing Manager/Payment Processor** – to complete the purchase transaction.
>
> **The Receivables Acceptor/Validator** – to ensure the supplier has delivered what has been purchased.

As in many of the case examples we studied, the sticking point so far has not been the technology, but the economics of making the scenario work profitably for both parties. The biggest adaptation here is that the machines can buy goods and services, not just communicate information. The commerce capabilities for Machines as Bound Customers begin to unlock the commercial potential of this new kind of customer. A shift that will advance even further in the next phase-Machines as Adaptable Customers.

Phase 2: Machines as Adaptable Customers

When machines become adaptable customers, they can choose from a wider variety of options to decide on behalf of the human -using a larger and more complex set of rules. The delegation of process steps is still largely human-driven because humans structure the process, and human intervention is still required for a few key steps. However, more aspects of decision-making are Machine Customer-owned, because AI gives them greater degrees of discretion and autonomy. When given the right data, a Machine Customer's AI can choose and act on behalf of the customer on select tasks with minimal human intervention. Adaptable Machine Customers have a greater ability to advocate a particular purchase or course of action than Bound Machine Customers.

Staples Easy System is a good example of an Adaptable Machine Customer. Introduced in 2016, the ordering system lets individual employees of its customer businesses order products and services from Staples by voice, email, mobile app, or text message. After an order is placed, the Staples Easy System consolidates purchases from multiple employees and devices into one shopping cart that can be approved by the appropriate office personnel. The Easy System also encourages the use of approved vendors and reduces redundant orders. The system also remembers customers' past orders to anticipate customer needs. If a customer requests blue pens, for example, the Easy System knows which brand of pens the office prefers based on the company's order history and orders that brand.[11] Given all the data this system must generate, it's not too wild to imagine that Staples' system will graduate to making product and service recommendations based on the customer's current needs.

Adaptable Machine Customers are useful in situations where substantial amounts of information need to be understood, multiple people participate in the purchasing process, and external factors like commodity prices and inventory availability fluctuate. For example, we know of a large software company developing a field service application for a luxury car manufacturer. The system can direct a car in need of repair to the right service station with the right parts and when the right technician is available. In consumer situations, Adaptable Machine Customers will be able to choose from a far wider set of variables than a human can when a decision needs to be made, but still within parameters set by humans. Variables could include price ceiling, location, and energy efficiency when buying a car or house.

Another example of an Adaptable Machine Customer is Genpact's GEPSmart. It is an AI-enabled digital inventory management system that can manage inventory across the supply chain in real-time while monitoring stock levels and supplier performance. [12] It can also send out restocking alerts, generate smart replenishment orders based on recommended or preset inventory levels, and forecast demand.

In these examples, the Machine Customer is working towards a goal set by humans, but they can choose between multiple options or methods to achieve the goal, just like human customers. The potential uses in business of Adaptable Machine Customers will grow as the machines' capabilities increase. Adaptable Machine Customers could function as more informed and empowered customers on behalf of their human owners. They could be active in more complex transactions such as contract negotiations or sourcing items from a wide range of marketplaces, freeing people to do more creative, higher-level work. We have even seen evidence of actual machine to machine (or AI agent to AI agent) communication and negotiation. In a widely seen YouTube video, two AI agents are seen negotiating details for an event at a hotel. At one point, one of the agents asks the other if they'd like to switch to a machine-based communication called Gibberlink. The other AI agent agreed and the rest of the video shows them conversing in their own language.[13] When machine-to-machine language and communication becomes seamless for machine customers, the possibilities for commerce are endless

To be able to select from multiple options (Table 7), adaptable custobots will need to have access to a larger and more diverse amount of data. They will also need advanced capabilities in decision management using machine learning coupled with a set of human-programmed rules "purchase this; if not available, then purchase that."

Table 7 - The Capabilities of Machines as Adaptable Customers

	Sense	Discover	Decide	Buy
Announcers	Sensors	Sensor Analyzer	Purpose Director	NA
Bound	Consumption Monitor	Needs Assessor Performance Manager	Purpose Director	Purchase and Payment Processor Receivables Acceptor Validator
Adaptable		Procurement Search	Trust Assessor	

Adaptable Machine Customers have several additional capabilities. **Procurement Search** enables the machine to search a wide variety of sources for information to inform its purchase decision. It is no longer confined within a walled garden. A **Trust Assessor** helps the machine know who to trust and who not to trust when it is considering suppliers. A **Peer Platform Communicator** enables the machine to communicate with multiple platforms.

Most of the capabilities that enable machines as announcers and bound and adaptable Machine Customers exist today in various degrees of maturity. But this will only take us so far. To truly capitalize on the trillions of dollars at stake in this ecosystem, a far greater leap in technology and human acceptance will need to happen: the functionality that enables them to act autonomously.

Phase 3: Machines as Autonomous Customers

In the next stage, things get really interesting: it's when machines can act autonomously and for themselves. Let us be clear before we jump in. We are not talking about Skynet in the Terminator movie franchise or Isaac Asimov's robots. We are talking about increasing levels of independent activity that, at times, mimic human customer behavior.

Autonomous custobots will be able to engage in complex decisions and purchasing activities with minimal guidance or direction from humans. The Machine Customer can act independently on behalf of human customers with a high degree of discretion and owns most of the process steps associated with shopping and buying. They will be able to negotiate and purchase products or services for consumers, corporations, or other machines. In many cases, Autonomous Machine Customers will be independent agents for a human or organization. In fact, a Machine Customer could be owned by or represented by multiple human owners.

The rise of agentic AI makes autonomous machine customers inevitable, which we'll talk more about later in this chapter. However, at the original writing of this book, examples were limited. In Chapter 2, we shared Tesla's ambitious plan to create a platform for its cars to work as robotaxis when not in use. In the meantime, other companies with robotaxi plans are already reported to be "letting them loose" on the streets of major cities.

- Baidu claims its robotaxis have a 10% market share in one part of Beijing[14]
- GM's Cruise robotaxis were taking public fare-paying rides in San Francisco. They have since decided to focus on autonomous systems for personal vehicles[15]
- Alphabet's Waymo is offering fully autonomous rides in Phoenix[16]
- Uber is working with UK software company Wayve.ai on autonomous taxis for London[17]

Once such free-range machines are out earning money for services without human intervention, we believe they will also be spending money as Machine Customers. That's because scaled-up robotaxi services won't be quite as efficient and economical if a human still must intervene to pay for gas, cleaning, or wiper blade replacement.

Are other Machines as Autonomous Customers in development as we speak? Most assuredly. Thilo Koslowski, the former CEO of Porsche Digital and now a board member of multiple automotive technology companies, believes autonomous vehicles will become their own legal entities.

> *"I mean, they could become their own business. You could be the investor. You buy one of those cars, register it to drive on roads, and then it makes so much money that it buys additional cars from its profits and becomes its own business entity. And it's a subsidiary of the person who originally purchased the first car or a group of people who 'invested' in the car at the beginning. There could be all kinds of different constellations to this. I do believe there's an opportunity for doing this, but somebody has to guide the cars' business. At the end of the*

day, the engineering going into machines will not infuse capitalistic principles. They will have to be programmed in such a way."[18]

Autonomous Machine Customers represent the most exciting aspect of the concept of Machine Customers: a brand-new customer opportunity. If these machines are free to choose their own supplier or service provider, they become a quantifiable segment of customers.

To help illustrate the potential of Autonomous Machine Customers, let us imagine an intelligent, autonomous beverage vending machine and its human owner, James.

Due to his other time commitments, James wants to turn over as much responsibility and decision-making to the vending machine as possible. As such, James trusts that its autonomous nature means that the intelligent vending machine will achieve its objectives without his needing to be intimately involved in the details. He places a high emphasis on the vending machine's efficiency, operation uptime, automation, and speed. His intelligent vending machine is entirely autonomous and can alert its supplier when it is out of stock or needs maintenance or repair. It can also forward the money it collects to its owner. Once it receives its owner's high-level objectives, the machine pursues them using a variety of levers such as variable pricing to stimulate demand, testing where products should be positioned to optimize sales, and communicating basic advertising pitches for the beverages it is trying to promote. It also automatically orders replacement drinks, negotiates pricing, finds the best supplier based on reviews, and more.

This connected vending machine accepts payments for purchases already and could have the ability to make its own payments for restocking or repair in the future. This would require banks to:

- Determine appropriate payment methods
- Offer a line of credit and cash management for the vending machine
- Market these services to the vending machine
- Validate the identity of the vending machine
- Execute a smart contract to finalize agreement with the vending machine
- Onboard and manage the vending machine as a small business customer

At this point, the intelligent vending machine becomes the customer to those that supply or service it. It is also a service provider to human customers who buy from it. This will mean other types of businesses – such as the drinks supplier, repair company, and even banks – must change their ways of doing business to accommodate it.

The key word is *proactive.* Autonomous custobots will default to proactive behaviors based on the stated needs and goals of their human owners, objectives that are general enough that they can make an independent judgment about how to best achieve those goals. For a supplier, Machine Customers will be tough customers:

- **They will be harder to negotiate with** because their motivations are well hidden in their internal rules and logic (and they all have perfect poker faces).
- **They can process large amounts of data** from a wide variety of inputs. They will carefully collect and weigh the data to make an informed decision based on logic and rules.
- **They don't need to be delighted**, so resources normally spent on exceeding human customer needs will be wasted.
- **They are not easy to persuade.** Machines may be more willing to lock into a supplier when the sales and fulfillment process works seamlessly and meets the requirements of the service level agreement. But if their programmed objectives change, it's *hasta la vista, baby*, even for a long-time incumbent.

These tough Machine Customers will be good for their owners:

- **They will be faster at processing data and have no problem repeating tasks.** You can give a machine a task once, and it doesn't forget to repeat it.
- **They will buy only what you need**. They will order exactly what the enterprise needs at precisely the right time and in the right quantity, which should have a positive impact on your sustainability and working capital.
- **They will find it easier to locate substitutable products.** It's not always easy to identify sourcing alternatives that are geographically closer with lower shipping costs – unless you read at light-speed.
- **They will request services before they are needed**. This will minimize downtime.

- **They will recommend value-added products.** They will be good at finding solutions that might be more expensive in the short run but cheaper in the long run (such as a service contract or warranty).

In cases where decision-making is largely rational, such as reordering chemicals, having a machine as a customer should be a great thing. However, if emotion plays a vital role in the value proposition or the product itself (in the case of luxury goods or other indulgences) or the project is very complex (custom-built IT projects, for example), Machine Customers may not arise at all.

In a Machine Customer world, marketing and selling will be data-science-based. Sales will be largely programmatic, and the process will be automated. Salespeople are still needed, but mostly for B2B and large accounts where it becomes essential to understand the human customers who are ultimately responsible for the purchase. Market researchers and data scientists can test, study and reverse-engineer machine behavior, looking for patterns that could inform their sales strategy. These roles will be analogous to the search engine optimization specialists we see today. An extensive effort will be invested in developing ways to manage and influence the logic and algorithms that machines use to determine their purchases. Marketing will also be programmatic and automated. Marketers will still be needed to understand the needs and behaviors of human end-customers, but they will also have to consider how machines have become integral parts of the customer journey. Let us examine how this may play out in a few situations in Tables 8-10.

Table 8 - Example: B2B Industrial Cleaning Services

Human Customer: Emotion-based marketing and selling	**Machine Customer: Programmatic-based marketing and selling**
• Professional image of employees • Cleanliness • Healthy environment • Safety • Price • Takes me out to lunch • Personable account manager	• Price • Location/Availability • Timeliness • Products employed and their specifications • Environmental impact • On-time performance • User reviews/ratings • Employee efficiency ratings • Statements of insurance and liability

Table 9 - Example: B2B or B2C Auto Parts Replacements

Human Customer: Emotion-based marketing and selling	**Machine Customer: Programmatic-based marketing and selling**
• Cost • Range of products • Fast delivery • Quick installation • Peace of mind • Brand name • Friendliness of salesperson or customer service	• Cost • Selection • User ratings • Engineering specifications • Shipping speed • Part availability • Part compatibility • Warranty terms • Country of origin • Environmental record of supplier

Table 10 - Example: B2C Toilet Paper

Human Customer: Emotion-based marketing and selling	**Machine Customer: Programmatic-based marketing and selling**
• Comfort • Reliability • Babies • Stuffed animals • Clouds • Durability • Value • Celebrity endorsers • Availability at my favorite stores/online	• Dispenser compatibility • Fragrance type • Enzyme/skin sensitivity • Hard water performance • Temperature range performance • Environmental residues • Environmental footprint/sourcing • Cost per foot • User ratings

You will notice that in all these situations, the machine will demand more information and not less to be able to decide. They will be rules-focused and logical. They will not be swayed by emotion or entertainment.

What Type of Machine Customer Will Be Right for You?

Well, it depends. As we have laid out, Machine Customer capabilities will increase with each phase of maturity. The more complex the buying situation and the more the human owner delegates, the more sophisticated the Machine Customer will need to become. For the time being, your use of Machine Customers will be situational – primarily Bound Machine Customers and Adaptable Machine Customers. That said, there may be a time when you have your own true and trusted virtual assistant, an Autonomous Machine Customer, managing most of your Machine Customer transactions. Your life could be coordinated by a kind of 'concierge custobot' that delegates to many other more specialized custobots. It's not hard to imagine something like Amazon's Alexa operating in that way via a substantial custobot ecosystem. You can't take a machine out to dinner, but you can feed it data. Aaron Rajan, Global VP of Consumer Digital Experience at Unilever, agrees:

> *"So, if you just look at the number of attributes that someone like Amazon requires to list a product, you'll see that there are many more attributes now to describe the products: all the sort of standard things from size, dimension, and weight, which are obviously related to supply chain."*[19]
>
> *"There are also consumer-relevant metrics, all the way through to those that have been sustainably sourced and ethically sourced,"* he said. *Yet as important as this is, there isn't an established set of metadata for attribution."*
>
> *"It is something that we spent a lot of time thinking about - what people care about to make sure we're building features that are going to be relevant for the values that people aspire to."* Rajan added. [20]

However, for all the opportunities that come with Autonomous Machine Customers, there are also risks. Autonomous custobots could:

Spawn multiple new security vulnerabilities and opportunities for fraud. This becomes critical given the billions of internet-connected machines. Untrustworthy organizations or machines may still be able to induce machines to purchase from them.

Complicate taxation. How is an Autonomous Machine Customer taxed? Is it considered a "person" the way a corporation is?

Raise complex legal questions. Who arbitrates a dispute when a machine as autonomous customer orders more fertilizer, but the owner/farmer does not want it once a truckload has been delivered?

Cost you friends. Human sales, service, and marketing people may resist the introduction of Autonomous Machine Customers that might risk their jobs.

Make unexpected demands on you. It's easy to think of Machine Customers as selfless silicon friends, cyber-dogs that look out for you and ask little in return. But in fact, they will have specific needs you will need to fulfill.

As you can see, the business opportunities created by machines as Autonomous Customers are equally interesting, risky, and thought-provoking. Even more, though shocking: Machine Customers will evolve to have their own needs in addition to the needs of their human owners. This will only be true for Machines as Autonomous Customers.

What Could a Machine Customer's Needs Be?

Machines will have a progression of needs and motivations as they go through the four-phase evolution to becoming Autonomous Machine Customers (see Figure 10.)

Figure 10 - The Hierarchy of Machine Customer Needs

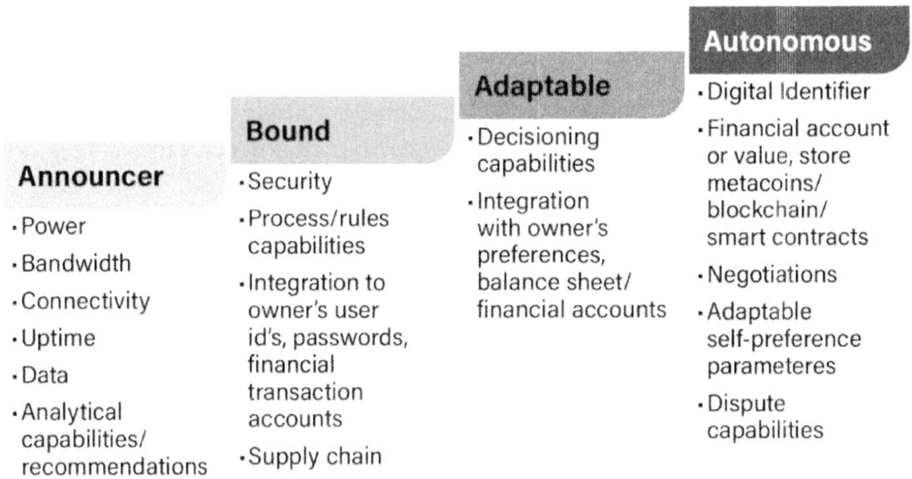

This Stair Step of Needs for Machine Customers include:

Machine as Announcer – Power, bandwidth, connectivity, uptime, data and analytical capabilities to make recommendations are basic needs of machines.

Bound Machine Customer – To process transactions, they will need security, business process and rules capabilities, and integration into the owner's user IDs, passwords, financial transaction accounts (like a credit card, debit card, or ACH account), and access to the company supply chain (if the owner is a corporation).

Adaptable Machine Customer – To make good decisions, they will need the ability to integrate the owner's preferences, balance sheet (accounts payable and accounts receivable), and financial accounts.

Autonomous Machine Customer – To make decisions independently of humans in this phase, they will need to have their own digital identity. They will need a government-issued identification to have their own bank account (given current laws in most countries) or cryptocurrency account.

What Are the Most Critical Technologies for Machine Customers?

Earlier, we outlined a set of capabilities needed for the various stages of Machine Customer development. With autonomous custobots, we don't need to add any new capabilities, but there are six technologies that will need to mature to bring them to life: IoT, Cloud/Edge Computing, Artificial Intelligence, Enterprise

Data Management, Decision Intelligence Platforms, and Blockchain. Figure 11 is a list of AI techniques, foundational, and enabling technologies needed to power Machine Customers.

Figure 11 – Machine Customer Technologies

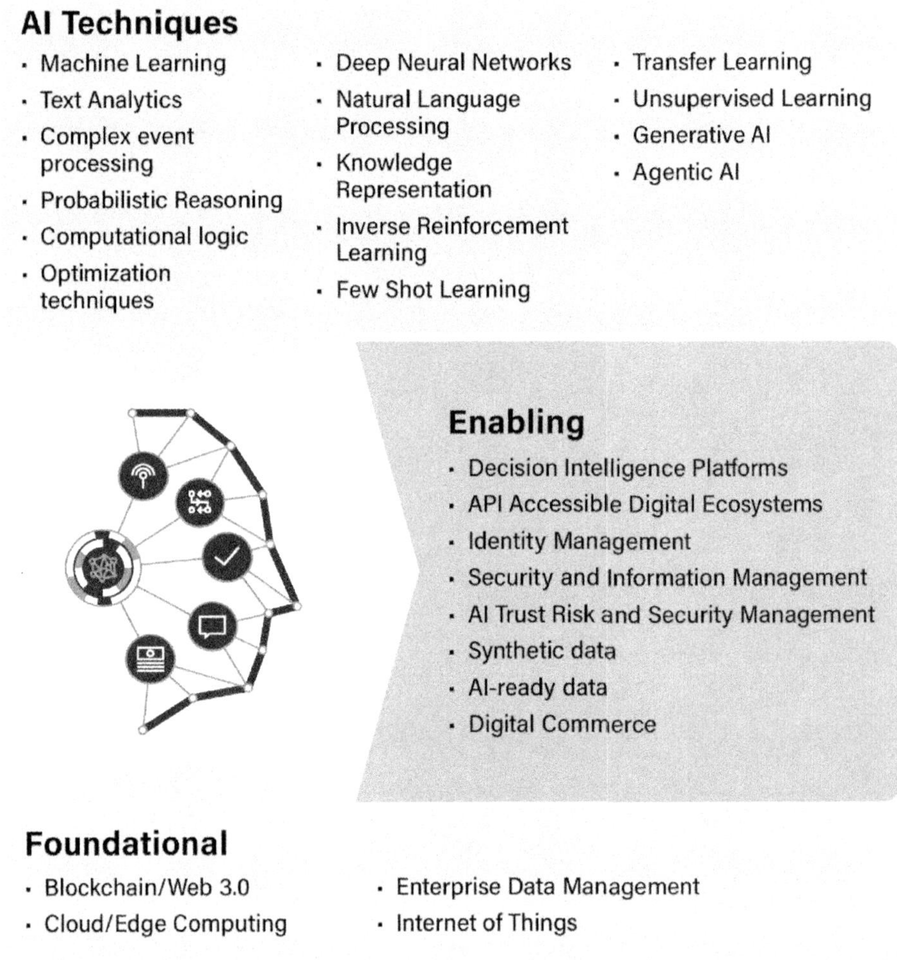

The authors freely admit this is largely a business book and not a technical book, however, for those of you who want to get a little geeky on technology, here we go:

The Internet of Things

An advanced Internet of Things will be supported by an array of technologies, most of which are several years away from mainstream adoption. The biggest is an edge-computing architecture that powers functionality as close to the device as possible to ensure fast data collection, integration with other systems, and real-time decision-making. However, implementing IoT is not easy, because of a lack of common standards and IoT development expertise. Geoff Parker, Professor and Director, Master of Engineering Management Program at Dartmouth College, told us:

> *"Things won't change until we go from hundreds of IoT platforms to a smaller number that has the scale to withstand nation state-sponsored cyberattacks."*

A European consortium of companies, the International Data Spaces Association, is working on that, according to Parker:

> *"They're trying to try to work out how do you watermark the data? How do you have contracts around it so that the machine level data can be sold to others and again, because of the massive scale of it,"* he said.[21]

Edge And Cloud Computing

Edge computing is important because it pushes as much computing power closer to the device itself as possible – in our case, the Machine Customer. It is defined as "a part of a distributed computing topology in which information processing is located close to the edge — where things and people produce or consume that information." Where is this "edge"? Well, the absolute edge is that place where the physical and digital world interacts, such as where sensors measure something happening in the physical world. The sensors then translate those observations or measurements into data, which can be used to decide to act, aggregated to look for patterns, or simply passed back to storage or analytics applications for further analysis.

For example, imagine a fast-food restaurant where the deep fryer, refrigerator, drink machines, and other critical equipment are connected and share their performance operating data or inventory levels. In a traditional cloud computing environment, that device-specific data would be sent to the cloud, processed, and sent back to the device or to an application. That takes time. With edge computing, smaller computing hubs (for example, those that run virtualized containers) in the restaurant can communicate with the devices and give store managers operating insight in real-time that they have never had before. Supplies can be ordered at any time because the computing power of a billion-dollar cloud system has descended to the kitchen.

Of course, the complement to the edge is the core or what is broadly known as cloud computing, that is, the upstream system or systems that guide what is passed down to the systems at the edge, often serving as the centralized storage, processing, and archiving "back end." Location-sensitive distributed computing is not a new concept; what is new is the explosive growth in the number of highly intelligent edge devices, coupled with the broad array of advanced services and massive scale of the cloud, serving as both orchestrator and core.[22] It is this network of computing that will give life to Machine Customers, moving them from isolated devices to nodes in a vast intelligent network with its own distinct set of possibilities.

Artificial Intelligence

If we compare a Machine Customer to a human, the device is like the body, while edge and cloud computing would represent the nervous system. Both are important, but it is artificial intelligence and advanced analytics that would constitute the brain. Artificial intelligence applies advanced analysis and logic-based techniques — including machine learning (ML) — to interpret events, support and automate decisions, and take action.[23] The key here is "take action." Together, machine learning and advanced analytics combine to give previously inert objects the chance to provide meaningful value.

Artificial intelligence is the magic wand that turns Pinocchio from a marionette into a real human boy. There is much confusion around the term AI. It has become shorthand for dozens of discrete technologies. AI is mainly a computer engineering discipline. This discipline is made up of software tools aimed at solving problems, not replicating the human brain (let alone the mind). From that perspective, the AI discipline (and toolbox) comprises a series of mathematical or logic-based techniques — uncovering, capturing, coding knowledge, and using sophisticated and clever mechanisms to solve problems.

Today, several categories of established techniques provide most use cases in AI.

Machine learning: Advanced machine learning algorithms are composed of many technologies (such as deep learning, neural networks, and natural language processing), used in unsupervised and supervised learning, that operate guided by lessons from existing information.

Text analytics is the process of deriving information from text sources. It is used for several purposes, such as summarization (what is the crux of this document?), sentiment analysis (what is the nature of commentary on an issue?), explicative (what is driving that commentary?), investigative (what are the particular cases of a specific issue?) and classification (what subject or what key content pieces does the text talk about?).

Complex-event processing (CEP) is a kind of computing in which incoming data about events is distilled into more useful, higher-level "complex" event data that provides insight into what is happening. CEP is event-driven because the computation is triggered by the receipt of event data. CEP is used for highly demanding, continuous-intelligence applications that enhance situation awareness and support real-time decisions.

Probabilistic reasoning: These techniques extract value from the large amount of data gathered by enterprises. This might include, for example, sifting through large numbers of customer records and identifying what the most important factors are for their satisfaction and how these factors are correlated.

Computational logic: Often referred to as rule-based systems, these techniques use and extend the implicit and explicit know-how of the organization. This technology will be central to managing the rules and logic a Machine Customer must follow.

Optimization techniques: Optimization techniques maximize benefits while managing business trade-offs, for example, constraint-based reasoning. They do this by finding optimal combinations of resources given several constraints in a certain amount of time. Machine Customers will be charged with making optimized decisions in many cases.

Deep neural networks (DNN) process data in complex ways by employing sophisticated math modeling. 'Deep' refers to a model's having many layers.

Natural language processing (NLP): NLP provides intuitive forms of communication between humans and systems. NLP includes computational linguistic techniques (symbolic and sub-symbolic) aimed at recognizing, parsing, interpreting, automatically tagging, translating, and generating (or summarizing) natural languages in spoken or written form. Machine Customers will need to interact with human actors.

Knowledge representation: Capabilities such as knowledge graphs or semantic networks aim to facilitate and accelerate the access to and analysis of data networks and graphs. Through their representations of knowledge, these mechanisms tend to be more intuitive for specific types of problems. For instance, new knowledge representations provide fertile grounds for AI techniques in situations where one needs to map out specific relationships among entities. Machine Customers will need to understand complex systems like markets and those that interact within them.

Inverse reinforcement learning is a method of imitation learning that aims to infer a reward function from expert demonstrations, rather than explicitly defining one. It's essentially learning from observing someone else do a task and figuring out what they were trying to accomplish. This approach avoids the need for manually specifying a reward function, which can be time-consuming and difficult, while still enabling the generalization power of reinforcement learning.[24]

Few-shot learning: Learning from limited datasets.

Transfer learning: Leveraging knowledge from one task to learn another.

Unsupervised learning: Learning from data without explicit guidance. These techniques will help machine customers learn from sparse data and reduce their reliance on large datasets.

Generative AI: Generative AI refers to artificial intelligence techniques that learn a representation of artifacts from data and use it to generate brand-new, completely original artifacts at scale, that preserve a likeness to the original data (including images, video, music, speech and text). This can enable the AI algorithm to improve or alter existing content, create new data elements and novel versions of real-world objects, such as buildings, parts, drugs, and materials.

ChatGPT, currently the best-known generative AI application, sits atop a large language model (LLM) called a generative pretrained transformer (GPT). LLMs are trained on billions of words and trillions of parameters. Complex math and enormous computing power are required to create these trained models, but they are, in essence, prediction algorithms - not unlike auto-complete on your smartphone. In the case of GPT and other LLMs, they predict patterns and sequences of words and generate language that makes sense in response to a question, called a "prompt."[25]

Tools like ChatGPT, Claude and will accelerate the development and deployment of machine customers. These applications can diagnose a job to be done and break down complex tasks to make the right recommendations, service requests and other functions necessary to complete the job. They can then create plans and even generate code to execute those plan steps. They will even be able to have a conversation with a human, such as the customer they are shopping for, or the customer service agent with whom they are negotiating.

Agentic AI

While we've explored how Generative AI forms the "brain" of Machine Customers, enabling them to process data and make decisions, Agentic AI represents a significant leap in machine customer capabilities. Agentic AI, or AI agents, are autonomous or semi-autonomous software entities that use AI techniques to perceive, make decisions, take actions, and achieve goals in their digital or physical environments. Until now, AI models, like Large Language Models (LLMs), have primarily performed tasks when prompted, such as generating text or summarizing documents. They lacked the ability to act on their own initiative.[26]

Agentic AI changes this fundamental limitation by giving AI new levels of *agency* – the ability to select what actions to take for achieving particular outcomes.[27] It introduces the concept of a goal-driven digital agent that autonomously makes plans and takes actions. This is not just automation triggered by specific inputs; agentic AI can receive high-level goals, iterate, delegate tasks, and provide dynamic information. It's AI taking initiative, adapting to context, and executing goals with minimal human supervision. To do this, Agentic AI will rely on Large Action Models (LAMs). These are advanced AI systems designed to identify and generate actions that can impact a target environment to achieve specific goals. They integrate pattern recognition with symbolic reasoning and can execute complex sequences of actions in dynamic environments. LAMs enhance the capabilities of machine customers by enabling them to perform complicated tasks that require a combination of reasoning and action execution, like the ability of a Machine Customer to search, select and buy items on behalf of human customers. Chatbots, for example, are increasingly popular agents.[28] Some Machine Customers will be software agents; others will be collections of agents.

The initial phases of Machine Customers involve being "Bound" to specific rules or suppliers and then becoming "Adaptable," able to choose among options but still largely following human-defined processes with required human intervention. Agentic AI is the technology that closes the gap to the "Autonomous" phase.

Additionally, agentic capabilities can enhance the other phases of Machine Customer development by going beyond simple tasks. They can examine data, perform research, compile tasks, and then take actions in the digital or physical world. Examples relevant to Machine Customers include analyzing order history, reviewing communications, setting follow-up actions based on behavior analysis, identifying upsell opportunities, composing personalized emails, and deciding on optimal timing. This will be just the beginning of the increasing sophistication and decision-making that Agentic AI enables in customer interactions.

Agentic AI represents a significant opportunity for performance gains. The future of decision-making will change as Agentic AI rapidly analyzes complex data, identifies patterns, and acts, leading to better problem-solving, reduced time to action, and enabling new concepts of scale. This increased speed and efficiency are critical for Machine Customers, who can operate in this new high speed exchange.

These techniques will form the foundation for Machine Customers. Agentic AI provides the key missing ingredient that allows AI systems to move beyond being passive responders to prompts and become active, goal-driven entities that can plan, decide, and act autonomously. Various combinations of the technologies will power the Sense, Discover, Decide, Buy model we presented earlier. When a Machine Customer is sourcing something for a human customer, these techniques allow it to sort through many different sources of information and events, interpret them in context, and then execute a series of actions to deliver on the desired outcome set by the human owner. Artificial Intelligence \make it possible for machines to automate and in some cases, mimic the behavior of human customers.

Machine customers can be viewed as a subset of AI agents. We can define the following categories of AI agents:

Goal-Oriented Agents. These agents act with a specific target state or outcome in mind, often optimizing for utility, cost, or effectiveness. Examples include: Autonomous Procurement bots and route planners.

Context-Aware Agents. These agents maintain an internal model of the world, reason about it, and plan actions accordingly. Examples include: Self-driving cars, Store Inventory Rebalancing Agents.

Optimization Agents. These agents evaluate different outcomes and select actions based on maximizing a utility function. Examples include: Dynamic pricing systems, Investment bots balancing risk and return.

Continuous Learning Agents. These adapt their behavior based on past experience, starting as simple agents but improving over time. Examples include: Game-playing AIs (like AlphaZero) and adaptive fraud detection systems

Organizations will have several options to develop, deploy and manage their AI Agents.

- **Packaged AI Agents**: These are ready-to-use AI solutions designed specifically for your industry and business needs. They allow your organization to quickly implement AI technology without the need for extensive customization, helping you stay competitive in a fast-changing market.

- **Agentic AI Platforms or Multi-Agent Systems**: These platforms enable your organization to create, manage, and expand AI solutions efficiently. They come with built-in features for coordinating AI tasks, learning from data in real-time, ensuring proper oversight, and maintaining data security, helping you leverage the latest advancements in AI and automation.
- **AI Agent Marketplaces**: These are online platforms where you can find and use AI solutions developed by various creators, including third-party developers and startups. Similar to app stores, these marketplaces encourage innovation and allow your organization to easily adopt and scale AI solutions tailored to your specific needs.

An emerging trend to watch is the concept of Guardian agents. Guardian agents are AI-based technologies designed to support trustworthy and secure AI operations. Guardian agents are a recognition that "human in the loop" processes aren't scalable with the proliferation of AI agents and moreover, individuals can induce their own biases and conventional thinking into evaluating the outcome of the agents. These agents are AI-based technologies designed to support trustworthy and secure AI operations. They function as both AI assistants (supporting users with tasks like content review, monitoring and analysis) and as evolving semi-autonomous or fully autonomous agents (capable of formulating and executing action plans as well as redirecting or blocking actions to align with predefined agent goals).[29]

Gartner AI experts tell us that AI agents are gaining traction in certain industries, using VC funding for startups as a proxy. Healthcare leads due to its paper-heavy, task-heavy nature, followed by financial services, transportation and logistics, and law/legal. Digital commerce and retail also have potential.[30] These are often complex industries, but the initial applications of AI agents within them are focused on narrow, simple tasks.

To reach autonomous capabilities, Machine Customers need to be able to develop abstract concepts and strategic plans from minimal data, such as goal instruction without explicit data inputs.[31] This results in producing contextual and dynamic output by iterating and delegating tasks, a capability ensured by embedding Agentic AI.

By 2029, it is estimated that over 50% of Machine Customers will reach the level of intelligence to be proactively autonomous, up from less than 10% in 2024, largely due to being embedded with Agentic AI capabilities.[32] This evolution in decision making will improve situational awareness and cross-domain knowledge via independent self-learning, expanding machine customers to newer domains. However, this autonomy requires advanced tools and strict guardrails for orchestration and governance, including legal and ethical guidelines on autonomy, liability, security, and data privacy.

In summary, Agentic AI is revolutionizing the landscape of Machine Customers by endowing them with the capability to act autonomously and intelligently, transforming them from passive tools into proactive partners. As organizations embrace these advanced AI agents, they will unlock unprecedented efficiencies and innovations, redefining how businesses interact with technology and their customers.

Data: Enterprise Data Management and Types of Data

The Machine Customer is fundamentally a data-driven entity. Unlike human customers, machines are entirely reliant on data and analytics to operate, decide, and learn. Their effectiveness, and the trust humans place in them, is directly proportional to the quality, accessibility, and breadth of the data they can gather and interpret. This urgent demand for data to support AI initiatives, including those powering Machine Customers, requires evolutionary changes to traditional data management. Simply put, AI-ready data is essential for AI success, and without an AI-ready data practice, over 60% of AI projects are predicted to fail by 2026.[33]

But what is "AI-ready" data? It is not traditional "high-quality" data, which often involves removing errors and outliers. It might seem counterintuitive to feed AI data with errors and statistical anomalies. However, it's shown that AI requires data that is representative, even including errors or unexpected but valid data points, for effective training of algorithms. It's a practice that cannot be built once for all data. Handling the data torrent generated and consumed by custobots requires significant investment in data science, AI, algorithms, and advanced analytics platforms. This means mastering techniques for interpreting events, supporting automated decisions, and taking action. An organization's data management practices must evolve to support AI-ready data, mastering skills in multimodal data management, retrieval augmented generation (RAG), unstructured data management, and using AI engineering platforms to ground internal data with AI models. As we have stated, Machine Customers operate via open API platforms and digital ecosystems, demanding seamless data exchange with partners and suppliers. Your existing digital commerce capabilities must be upgraded to handle these machine-to-machine requests and the immense data volume. Foundational data management technologies like data integration, observability, metadata management, and governance remain essential components of an AI-ready practice.

In some cases, you may need synthetic data to help train and manage your Machine Customers. Synthetic data is a class of data that is artificially generated rather than obtained from direct observations of the real world.[34] It can be created using various methods, including statistically rigorous sampling from real data or generative approaches such as generative adversarial networks (GANs) and foundation models. An emerging technique is the use of advanced simulation in a virtual environment to simulate scenarios and create entirely new datasets of events. Unlike real-world data, which often contains biases, is incomplete, or noisy, synthetic data offers the ability to generate data that is desired but unavailable or difficult to obtain. It has been used for years to augment real data.

Synthetic data is crucial for the development and advancement of AI agents because it empowers AI by overcoming many limitations of real enterprise data. The availability of fit-for-purpose and compliant synthetic data fuels AI innovation, driving domain-specific solutions. As the need for more data grows, the synthetic alternative provides a scalable solution to generate diverse and high-quality datasets, thereby improving model accuracy and reducing biases. It is also essential for system testing and creating realistic product demos for AI systems. Furthermore, synthetic data allows organizations to work with data that mimics real-world scenarios without exposing sensitive information, addressing critical privacy concerns and regulatory compliance requirements for deploying AI solutions. It enables the exploration of extreme and rare scenarios through advanced simulations, which is particularly vital for training machine customers in complex sectors like automotive, healthcare, and finance.

Beyond enabling data flow and preparing synthetic or AI-ready data for Machine Customers, managing this data is a complex and critical undertaking. Traditional data governance frameworks must be adapted for machine entities, requiring a flexible governance model that can balance centralized and distributed approaches based on the AI use case. Data quality, accuracy, and timeliness are fundamental requirements. Metadata management is really important, especially when dealing with complex data types and data that's used on-the-fly. It helps keep track of where data comes from, makes sure the data is understood correctly, and allows for better control and cost management.

But enabling data flow is only the first step; managing this data is a complex, critical undertaking. Existing data governance frameworks, built for human-centric processes, must be adapted for these new machine entities. Data quality, accuracy, and timeliness become non-negotiable, as interconnected machines become prime targets for tampering and data breaches, demanding new systems to detect anomalous activity unique to Machine Customers. Furthermore, the ethical dimension of data looms large. Bias in training datasets can subtly warp algorithmic decisions, creating opportunities for exploitation unless digital ethics are hard-wired to ensure machines act in the beneficial interests of its human user.

Decision Intellgence Platforms

Decision Intelligence Platforms (DIPs) are software designed to engineer decision making, supporting, augmenting, or automating decisions for humans and machines.[35] They achieve this by combining explicit decision modeling, AI, analytics, and related capabilities. DIPs are an evolution of earlier decision support systems, focusing on decision-centric solutions to move from data-to-insight to insight-to-action, acting as a bridge. They allow the composition of various techniques like machine learning, computational logic, and AI agents into cohesive solutions centered on optimizing specific decisions. More advanced DIPs incorporate techniques such as AI agents, simulation, and GenAI to enable situational and adaptive decision flows, boosting autonomy in operations.

This technology is crucial for Machine Customers as they rely on AI and advanced analytics as their "brain" to sense, discover, decide, and buy. As Machine Customers evolve through Bound and Adaptable phases towards full Autonomy, their decision-making becomes increasingly complex, requiring them to process extensive data, identify options, and decide based on varied parameters. DIPs provide the essential platform to explicitly model, orchestrate, and execute these sophisticated decision processes, powering the core functions required for Autonomous Machine Customers to operate effectively in markets.

Blockchain Technologies

A blockchain is an expanding list of cryptographically signed, irrevocable transactional records duplicated in every node of the network. Each record contains a timestamp and reference links to previous transactions. With this information, anyone with access rights can trace back a transactional event, at any point in its history, belonging to any participant. A blockchain is one architectural design of the broader concept of distributed ledgers. Smart contracts are one example of what can be enabled by Blockchain.

For machines to be customers, they will need to have some type of purchase agreement. In a world that makes increasing use of the security of blockchain technologies, it would be called a "smart contract"[36]. This is a type of blockchain based self-executing contract with the terms of the agreement between buyer and seller being written software. From opening the door to buying a large piece of equipment, these smart contracts operate in real-time, and they are runtime code. Will all Machine Customers operate with smart contracts? We think not for routine B2C transactions. However, for B2B transactions where contracts are commonplace, having an encrypted and unalterable record of a transaction would establish a Machine Customer's credibility very quickly.

What Are the Other Digital Enablers Needed for Machine Customers?

Technologies that enable the simple functioning of a Machine Customer are important. But once the Machine Customer moves beyond the Bound phase and can choose from multiple suppliers and make decisions, other enablers must be in place. The key ones include:

- API accessible Digital Ecosystems and Marketplaces – such as Alibaba's Taobao and Amazon Marketplace
- Identity Management – such as Access Management and Authentication and Identity Verification
- Security technologies - such as Network and Endpoint security, Identity Access Management and Data Protection
- AI-specific Trust, Risk, and Security technologies

You can also think of these enablers as being central to maintaining the health and vitality of the Machine Customer. Will the Machine Customer have access to what it needs to fulfill its programming? Are the technologies secure enough to protect the Machine Customer and the human owner from harm?

Platforms And Ecosystems

A platform is a common substrate that serves or enables multiple products or services. Platforms (in the context of digital business) exist at many levels. They range from high-level platforms that enable a platform business model (think Amazon) to low-level platforms that provide a collection of business and/or technology capabilities that other products or services consume to deliver their own business capabilities (for example, Airbus' Skywise platform). Platforms that enable a platform business model by nature have a business ecosystem attached to them. For example, Alibaba is a digital commerce multi-platform business that matches millions of customers with millions of products and suppliers. Alibaba also has adjacent products like 1688.com, (a B2B trading platform in China) Taobao (a C2C and B2C trading platform in China), TmallGlobal (a B2C International platform into China), and AliExpress (a B2C China platform into international markets). These platforms typically expose their capabilities to members of those ecosystems via APIs or application program interfaces – a digital connection for software programs to access rather than humans.

Platforms are important to the utility of Machine Customers. In working with platforms offered by digital giants like Alibaba, Amazon, Apple, Google, and Microsoft, and Salesforce, businesses can ensure their products and services are exposed to the greatest number of customers. And they avoid the expense of building their own complex software to serve Machine Customers. Most businesses are likely to use this approach.

While the initial, simpler Machine Customers might operate within vendor-specific "walled gardens," achieving widespread adoption requires neutral platforms or market spaces where machines can transact openly. The challenge to scaling true autonomous AI agents acting as Machine Customers includes the very platforms required to orchestrate potentially thousands of these independent entities.

iProd, an Italy-based software company, gives us a glimpse of this future. Their platform collects, manages, and optimizes the key operational areas that determine the efficiency of a manufacturing company including preventive maintenance. Stefano Linari, the founder of iProd, claims iProdMOP as the first platform that can accommodate the requests of Machine Customers located in a manufacturing environment connecting them with their manufacturers and even with consumables resellers hosted in iProd IoT marketplace.[37]

Identity Access Management (IAM) for Machine Customers

At a Gartner conference in late 2024, Don and Mark led a roundtable discussion on Machine Customers and something interesting happened. Up until that point, these discussions were about explaining the concept of machine customers and getting clients to see how they might fit in their business or organization. Instead, we spent most of our time discussing the identity of machines and how to guard against fraud. From our perspective, this was a turning point in the acceptance of Machine Customers. New questions arose: How do you verify the identity of an AI Agent acting as a Machine Customer? How can that Machine Customer verify your identity? As we said earlier, the entrance of payment processors like Mastercard and Visa will play a large role in identity verification as they do for today's human-based transactions.

The importance of a well-defined identity for a Machine Customer, particularly an AI agent acting autonomously, cannot be overstated.[38] Having an identity is essential for authentication and authorization, allowing the machine to securely access applications, data, or other resources, in line with zero-trust principles. IAM initiatives aim to ensure that machines, alongside humans, access the right assets at the right level, time, and for the right reasons. For Machine Customers to act, they must be able to reliably prove who or what they claim to be. For Autonomous Machine Customers to operate and engage in complex purchasing activities, they will need their own digital identity to make decisions and potentially even have their own bank accounts or cryptocurrency accounts, given current laws. Effective management and governance of these identities are critical. Without them, organizations face significant risks, including unauthorized access, data breaches, and operational disruptions, which undermine trust and compliance in critical systems. Ultimately, for economies to benefit from Machine Customers, these machines may eventually need to be granted their own legal personhood, even on a constrained basis, requiring trust and accountability framed by their rights and responsibilities.

Establishing an identity for an AI agent acting as a Machine Customer involves several key elements:

Identifiers: The identity must be labeled with one or more unique attributes that distinguish it from other machines within a domain.

Attributes: These describe the individual traits of the specific AI agent, providing context within the systems it operates.

Credentials: Authentication data, such as certificates, JWTs, or cryptographic keys, are associated with the identity to provide trust when the AI agent claims that identity during a digital interaction.

Technical Capabilities: A machine IAM program requires capabilities like discovery to catalogue existing secrets and credentials, life cycle management, issuance of credentials, access control based on defined policies, and continuous monitoring.

Dynamic Service Identities: For many interactions, particularly API calls, instead of relying on manually created accounts, the AI agent might authenticate using temporary credentials governed by dynamically created and automatically managed service identities controlled by policies, such as requiring a token with specific claims or a recognized certificate.

Security Information and Event Management (SEIM)

The last of our enabling technologies centers around trust and security. As Machine Customers are at risk of being hacked or manipulated, causing intentional or unintentional harm, their security is very important. While there are many types of device security and security software services, we felt the one to focus on is security information and event management (SIEM). This technology supports threat detection, compliance, and security incident management through the collection and analysis (both near real-time and historical) of security events, as well as a wide variety of other event and contextual data sources. The core capabilities are a broad scope of log event collection and management, the ability to analyze log events and other data across disparate sources, and operational capabilities (such as incident management, dashboards, and reporting).

Security is important to societal acceptance of Machine Customers because if they are not secured, it will be nearly impossible to trust them to conduct transactions on behalf of a person. We expect that the privacy and security laws that apply to people around the world will eventually be amended to include Machine Customers that act on behalf of a person. Geoff Parker, Professor of Engineering at Dartmouth College, believes that security will be critical to maintaining relationships between organizations in a Machine Customer ecosystem:

> *"The angle is really one of knowing how vulnerable these systems will make us when we interconnect them all. It's going to require scale. Then, once you have that scale, you can migrate across the different industry participants. So, it's almost as though the security side is going to drive a scale that will then make cross-industry cooperation that much easier."*[39]

AI TRiSM: Ensuring Trust, Risk, and Security for Machine Customers

We have discussed how Machine Customers function, heavily relying on artificial intelligence and other advanced technologies. What comes next is preparing these new systems to be safe, reliable, and ethically operated. This is where Gartner's AI TRiSM framework becomes essential. AI TRiSM, or AI Trust, Risk, and Security Management, is the technology stack that supports the implementation and enforcement of enterprise AI governance policies. It provides a structured approach as organizations deploy and operate their new AI systems. Considering Machine Customers are powered by AI, this framework is directly applicable to governing their behavior and ensuring they operate safely and effectively within your business ecosystem. It outlines key technology functions necessary to monitor, validate, and manage the risks associated with AI-driven agents.

This includes:

AI Information Governance Technology: This technology supports the governance of data and information used by AI systems. It has an initial focus on data protection, data classification, and access management. This function is crucial because Machine Customers are "astonishingly data-thirsty". This required data will need to be managed for the diverse AI and IoT systems your organization might employ.

AI System Discovery and Inventory Technology: This function enables organizations to discover and inventory all AI systems and tools used in the organization. This is a foundational step for establishing governance and understanding the AI landscape.

Guardian Agents: Guardian agents are AI-based technologies designed to support trustworthy and secure AI operation. [40] They function as both AI assistants (supporting users with tasks like content review, monitoring and analysis) and as evolving semi-autonomous or fully autonomous agents (capable of formulating and executing action plans as well as redirecting or blocking actions to align with predefined agent goals). These controls are necessary for managing the risks associated with AI, including security vulnerabilities and opportunities for fraud.

Conclusion

As Machine Customers become more autonomous from their human owners, the complexity of the opportunities, risks, and technologies will rise exponentially. Machine Customers will have their own set of needs, which will be very different from those of human customers. Owning a Machine Customer will require changes to your business operating model. And if you sell to a Machine Customer, sales and marketing practices will become less human-based and more programmatic. Like it or not, Machine Customers are already emerging, and we must find a way to integrate them into our lives, our organizations, and our connected society.

Key Takeaways

Machine Customers exist today and are developing.

From announcing information to buying from multiple suppliers, machines are taking on more and more human work in the purchasing process.

Action: Consider the way your human customers buy from you. Ask yourself, could a Machine Customer take over for your human customers? How would your organization respond to that shift in the areas of people, process, and technology?

Machine Customers require a different set of capabilities.

These will include extensive use of data, analytics, and AI.

Action: Reassess your mid-term talent strategy in these areas. You might need to hire more data scientists and fewer sales and marketing people.

Emerging technology and digital ecosystems enable Machine Customer autonomy.

Here we reach the final frontier of sales and marketing: machines as Autonomous Customers.

Action: Make sure your capabilities in digital commerce, digital ecosystem platforms, blockchain technologies, and security are current.

Chapter 8

How Machine Customers Could Hurt Your Business

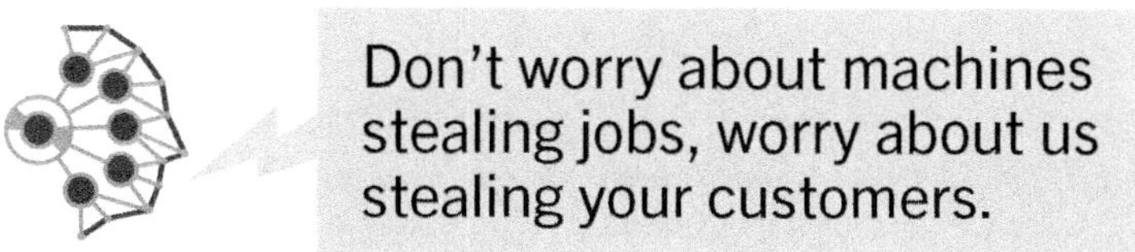

Any market transition is a double-edged sword with big winners and losers - muddling through this one won't be an option

Machine Customers Create an Opportunity for Market Disruption

Most businesses operate in mature markets, stuck on a slow steep treadmill to drive consumption while limiting customer churn and replacement. A big new market is a rare and exciting opportunity that people will remember for decades. Back in 1996, for instance, only 16% of adults had a cell phone. By 2016, 95% did.[1] Today, the glory years are over, and the cell phone industry has reached saturation. Providers are locked in a constant battle to retain share, thwart commoditization, and maintain margins – and like most other mature businesses, keeping an eye out for fresh wide-open markets somewhere else.

Machine Customers will create substantial open market growth opportunities in three ways. First, they will add to the total number of human customers by taking away customer 'work' in signing up for a new service. Often, the overhead involved in participating in a market effectively reduces its size. For example, the uninsured, the unbanked, and the unfit fail to buy in part because of the friction they think it will involve – too many decisions, too many contracts and too many forms. We think Machine Customers can help with that. Here's a hypothetical dialogue of the future between a consumer and their smart speaker-based AI assistant.

"AI Assistant – I think I need to start going to a gym for my New Year's resolution".

"Sure Jules – there's one suitable for you as a beginner, without a contract lock-in less than two miles away. To pay for it, may I cancel that third streaming TV service you haven't watched since August?"

"OK, AI"

"Jules, the smart mirror often sees your training shoes, but we don't think you own any gym wear. Unless you decline, I'll buy three sets of basic shorts and tops tomorrow to get you started."

"OK, AI"

Second, Machine Customers will increase total sales in many markets by being more efficient and timely purchasers. Think of all the seasonal purchases from ski holidays to lawn seed that we don't quite get around to in time. Think of all the occasions when you realize you need something – from garlic granules for your recipe tonight to a guest presenter for your January sales kick-off meeting – and found it was too late to get them.

Third, custobots will start to become customers for themselves. This is when the biggest open field opportunities will arise. For example, right now, drone and droid e-commerce delivery services are few and tiny in scale. A decade from now – they will be massive and everywhere. Who will serve all those drones and droids when they get stranded, need a replacement part, or want a clean? Everything can become a smart machine. Every smart machine will start to have its own needs. The smartest machines will even start to earn their own money and learn to support themselves.

This is where the biggest open-field opportunities will arise. As machines evolve to become customers for themselves, entirely new markets and market locations will be created. Machines can go places people can't or won't. They will do work people can't or won't. They will clean more toilets and clear more gutters. They will offer robotaxi services in less populated towns, not just big cities. Their cheap CubeSat cameras and sensors will stare down, detect and report activity with ceaseless vigilance from low earth orbit, increasing consumption of energy, cleaning fluid, insurance, and bandwidth even as they make crops grow better and factories run more efficiently. If the machine is in a place that people won't go, doing things that people won't do – why would we imagine a human is necessary to act as a buyer on its behalf? That makes no sense. Having the machine buy for itself is the better option and often may be the only option.

In many industries gaining or retaining a reasonable amount of market share, control is a constant battle. Without sufficient control over how a market operates – sales and margin predictability are lower, volatility higher, and the management team finds it tougher to win. Gaining control over a market – in the sense of being a leading market player – can be achieved in many ways: cost and price, product innovation and features, brand, platform, standards, asset exclusivity, or a regulatory moat. However, in many situations, there are no new moves to be made; all the permutations have been tried. The arrival of Machine Customers opens a brand-new way to gain market control that has not been available before.

Many industries are owned by powerful but lazy oligopolies. Mid-size and smaller players looking to disrupt these too-cozy clubs might use Machine Customers as a way in. The first mover that introduces Machine Customers to a previously human market could try a classic Clayton Christensen disruption. Christensen's famous innovator's dilemma[2] describes the paradox of success in mature markets. Incumbents keep adding features to products, bloating them to the point where prices are high, and submarkets underserved. Complexity becomes the enemy of total market size. The few players bicker over the portion of the market that is prepared to pay for a fully featured product, while making little or no effort to reach potential customers with a cheaper stripped-down version. This leaves space for a disrupter, who enters and grows via the unserved consumers neglected by the incumbents.

Machine Customers could be a powerful variation of the traditional Christensen gambit. That's because while Machine Customers might *appear* to be highly complex, they will be market simplifiers. New entrants may introduce a stripped-down product into the market, targeting Machine Customers.

Incumbents often evolve overelaborated and overpriced propositions that prey on human customer weaknesses and introduce artificial differentiation – all supported by additional marketing costs. Machine Customers will tend to cut through the illogical noise, ignore low-value features and seek minimal but sufficient solutions. The product they select will be cheaper (partly because it is streamlined and partly because of lower marketing and sales costs), and this will help to grow the disrupter's addressable market, almost out of sight of the incumbents.

For example, table salt is a simple product, but that hasn't stopped marketers from using branding, packaging, and merchandising to win shopper's loyalty. When Machine Customers start buying salt for their owners, they will quickly realize there is little or no functional difference between the heavily marketed, high-priced brand and the generic supermarket label. What will the brand owner do? The typical response will be to add more features. The brand might begin to promote many qualities that humans could not be bothered to read about on the label and use in decision-making, such as information about consistent average grain size, traceability of source supply, or pop-science health additives. However, that ploy won't work on custobots. They will learn to ignore and filter out added irrelevancies far more quickly than more gullible human customers.

Machine Customers Will Also Create Headaches for Your Company

Serving Machine Customers won't be easy. First, businesses must undergo a lot of adaptation and adjustment that will involve significant change costs. Second, they may have to deal with a "more intelligent" Machine Customer that reduces the total market size.

Machine Customers will tend to behave more rationally and operate more quickly. The good news is that it might help reduce brand marketing costs and shorten buying cycles that are elongated by human dithering. Machine Customer markets might also usher in lower transaction costs and higher distribution efficiencies at significant risk to many incumbents. Efficiency always sounds like a good thing, doesn't it? But many businesses' margins depend on inefficiency, overbuying, and waste. This is even true for B2B companies. Sherry Aaholm, Vice President and Chief Digital Officer at Cummins Inc., describes the balance between efficiency and cost this way:

> *"Right now, we're doing testing activities on engine NOx sensors. We're trying to predict the failure of a NOx sensor for end customers and proactively fix the sensor when the engine comes in for normal maintenance work instead of repairing it unexpectedly and causing unplanned downtime for the customer. But the challenge in that particular innovation is even though we have a very promising data model, our business still requires us to ask the question, well, wait a minute, what if our data model is wrong,*

and we end up replacing the NOx sensors and they were not going to break? Then we'd be absorbing a cost that we wouldn't have had to pay for otherwise. So, what I'm having to do in that case is to actually demonstrate for a period of time, the success of that model, before people are willing to give up that risk and willingly trust the data. It is doable and we will be one of the first in our space to bring it to market, but it requires changes in our business processes and more than an accurate algorithm."[3]

A good everyday example of this is fast fashion, many millions of tons of which end up in landfills. Another example is food. For example, in the UK, 9.5 million tons of food go to waste every year, and 70% of that loss takes place within households.[4] Who can be surprised when everyday shoppers are tempted by "two-for-one" offers into buying more than they will consume? Parents capitulate to child pester power and buy unwanted volumes of breakfast cereal because of the toy offers on the pack. "Use-by" dates prey on safety concerns causing consumers to throw away perfectly good food. Marketers exploit our preference for immediate rewards ("hyperbolic discounting" in economist-speak) by linking buying goods and services to experiencing pleasure or avoiding negative emotions.

Mechanisms like this can inflate the total size of a market, often substantially. Why do you think you need a new car every two or three years? Perhaps it is partly because of all those beautiful ads showcasing the latest design alloy wheels, blowing through autumn leaves in slow motion on hairpin bends in empty rolling-hill landscapes. You might think that you are too smart to be influenced that way. OK, will you accept that other people are? If not, then the only alternative left is that all car advertising is a waste of money, and that seems unlikely. Through that advertising, attractive leasing offers and the suggestion that personal social status is partly defined by what you drive - a major market is inflated in total size.

Companies and even whole industries sometimes get into a dependency habit, maintaining revenue by holding onto customers in sly and surreptitious ways. This ill-gotten revenue is often facilitated by digital commerce and customer service. In his book *Deceptive Patterns: Exposing the Tricks Tech Companies Use to Control You*[5], Dr. Harry Brignull describes a series of "tricks used in websites and apps that make you do things that you didn't mean to, like buying or signing up for something". They are also used to inhibit you from doing things they don't want you to do. But Machine Customers could be a force that disrupts the use of these methods, undermining the revenue that flows from them. As Joshua

Browder, CEO of DO NOT PAY, tweeted: *"A.I. will end dark patterns, because it doesn't give up as easily as a human!"*[6] – right after he successfully deployed the company's generative AI Custobot to negotiate with the New York Times to cancel his subscription.

Machine Customers won't fall for the twofers, the fluorescent packaging colors, the Instagram social peer pressure, or the cleverly worded fine print hidden in the financing offers. If they insert their penetrating, dispassionate logicality into a market - businesses that rely on waste could face a severe reckoning.

Buying less unnecessary stuff is obviously good for the environment, which means that the planet and society are likely to benefit from Machine Customers. However, some industries may have to go through the pain of major shrinkage.

When machines become customers, they act much faster than humans. Often their needs can only be met by other machines. Andrea Ciccolini of Amplifon sees the future of Machine Customer marketplaces this way:

> *"I envision very few one-on-one channels, and more many to many, according to a framework that certifies and create segments and clusters of machines for which the interaction may be more fluid, more secure in and for which there is a recognition process built into it."*[7]

However, efficiency can have its own downsides. A bot-to-bot marketplace can also go wrong much faster than a human market ever could.

The weather was quite fine on May 6, 2010, and market traders in New York were upbeat about the news of a $1 trillion plan from the European Union and the International Monetary Fund to avoid a European debt crisis, when suddenly, a little before 3 o'clock (Figure 12):

Figure 12 - A 'Flash Crash' Chart for the S&P 500 Stock Index

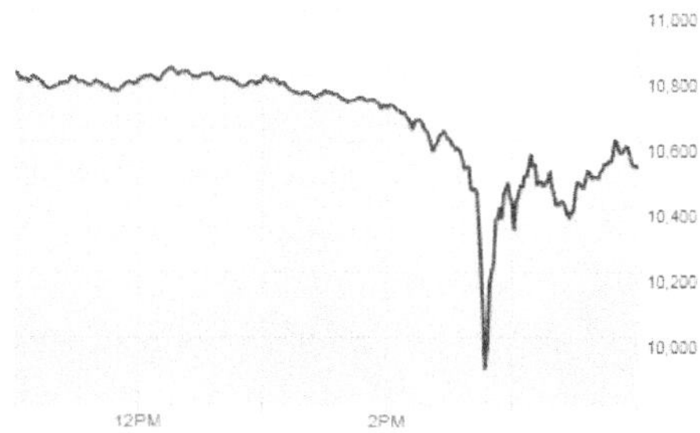

Stock indices, including the S&P 500, Dow Jones Industrial Average, and Nasdaq Composite, all fell about 9% within a few minutes. Though they rebounded very rapidly, it was one of the biggest intraday declines ever seen. SEC investigators later identified multiple causes, one of which was automated high-frequency trading. "HFTs began to quickly buy and then resell contracts to each other – generating a 'hot-potato' volume effect as the same positions were rapidly passed back and forth."[8]

Something similar could happen in other industries when goods are sold by machine to machine. For example, imagine that one day Adele posts a photo on Instagram, and in her hands, on the cover of the vinyl of her brand-new music album, is a picture of her wearing a new style of designer jeans.

Social media spots the jeans. AI intelligent assistant bots notice that their consumers will almost certainly enjoy the jeans that have been liked by friends in their personal networks. The bots start scouring the market for a similar style using image recognition, find it, and order it preemptively using the authority their human customer gave them. There isn't much risk for the customer: retailers will take returns, and the credit card companies back the transaction if the merchant does not deliver – so the bot's action is safe enough to be common practice.

All those purchases by the machine mob mean that brands that make similar styles of jeans run out in minutes. The algorithms in control of apparel manufacturers with autonomous operating models sense the shift and start reprogramming production lines to produce more of the new style. The price premium on the new style means that some companies decide it's worth switching production. Existing contracts for other styles are held back or canceled. Smart contract clauses kick in, and penalty fees are automatically applied, but the margin lift covers that. The whole jeans-making and distribution ecosystem pivots in a matter of 48 hours.

A couple of days later, journalists digging into the past of Adele's little-known designer discover an unsavory personal incident and expose it. The new jeans design trend is canceled overnight. The personal shopping bots rescind their orders *en masse.* Jeans in flight become returns, jamming warehouses. Manufacturers lose the projected margin gain immediately but must still pay the penalties on the contracts they broke. Smaller retailers quickly run into cash flow troubles and don't immediately refund the custobots for the price of the jeans they ordered. Credit card companies take a hit for the merchant's failure. Thousands of tons of denim products end up going straight to landfills or incinerators, incurring waste handling charges and carbon credit costs. Everyone involved in the whole ecosystem loses, and it all happens so fast, some players don't even know what hit them.

Such a scenario is possible if we do not develop the Machine Customer future thoughtfully and learn from our mistakes as it evolves. For example, one of the systemic fixes applied in the aftermath of the 2010 stock market flash crash was the addition of 'circuit breakers.' If a market falls too far, too fast, trading is automatically suspended for a period. If it happens again when the market restarts, then the next period of suspension is longer. Maybe fashion will need a similar mechanism.

Machine Customers' Hyper-Efficiency Could Be Commoditizing

Machine Customers are also likely to network, share and collaborate well. Combined buying power is likely to lead to commoditization. Commoditization lends itself to consolidation. The dispassionate, measurement-obsessed focus of Machine Customers will be ruthlessly efficient at evaluating products and services. Market variations of products and services will be compared. Machine Customer markets will tend to converge on the same optimal solution – the same best version of a product or service design. Human markets do this, of course, but Machine Customers will lead to faster and deeper commoditization.

Machine Customers are likely to be better at price negotiation than humans. Their endless vigilance, patience, and typically superior math and game logic will win out. If a better price is available because of the time of day or batch size – they will spot it and home in on it. And because they are always on and always connected, they will be able to network, collaborate and swarm. Tactical buying power aggregation will be easy and natural. All these factors will tend to push margins down, giving producers little financial headroom for differentiation. Again, the commoditizing force will be powerful. These factors can already be seen in the way online human customer markets behave. Machine Customers will take this behavior to a new level.

Highly commoditized markets - from cement to pork bellies - have always tended to consolidate. Machine Customers are likely to increase commoditization, driving markets further towards oligopoly or even last man-standing situations. Regulators may find they must control the power of Machine Customers in some markets for fear that short–term efficiency excess and overreach on the customer side could make the few remaining providers unsustainable in the mid–term.

And There is an Even Darker Side

Machine Customers are a powerful tool and should be treated with caution. If your organization uses or serves Machine Customers, you must take measures to manage risk in the event Machine Customers are influenced or duped by bad behavior – based on policy, humans making bad choices or AI, or some combination of the three.

For example, it might be tempting to make one small change to a Machine Customer to affect a business outcome. This might be by tweaking the re-order point for a consumable earlier in the purchase cycle or recommending maintenance off-cycle to meet a revenue target. While there may be a short-term benefit to this type of manipulation, in the long term it will make humans wary of Machine Customers. Three scenarios worry us most:

Scenario 1: Operating Failures

A bot controlled by a human performs an action that has an unexpected consequence for the Machine Customer, leading to some type of damage.

In this scenario, physical assets themselves get damaged temporarily or damage things surrounding them. In most cases, the damage ranges from a cold home to substantial trading losses on the stock market. That said, lives are not lost, and operations get back to normal relatively quickly. But it could be worse. We have already seen:

Autonomous vehicles crash. Autonomous vehicles (AVs) will coexist on the roads with human drivers for a long time. This makes their implementation an extremely complicated task, as AVs will have to drive safely along with humans, whose driving patterns are unpredictable along with road conditions they may not have had time to be trained on[9].

Cameras in your devices that measure your reactions. In 2011 Verizon filed a patent to track the behavior of TV watchers as they sing songs, play with a pet, or enjoy some supposedly private time with a loved one on the couch.[10] The tracking system would then search terms related to the behaviors it sees — such as "cuddling" or "romance" — and present viewers with TV ads related to that topic during commercial breaks, according to the patent filing.

High-speed robotrading run amok. On February 6, 2018, the U.S. stock market plunged 700 points in 20 minutes and 1,175 points for the day. The U.S. Treasury secretary said algorithmic trading "definitely had an impact".[11]

Software updates that compromise a critical system. Outages of Google's Nest wi-fi-enabled thermostat are reported sometimes. Once, for example, a bug in a software update led to draining battery life of thermostats and deactivated the system, leaving homeowners unable to heat their homes[12].

As the number and reach of Machine Customers increases, we expect that the scale of business and people affected will increase. For example:

Demand for autonomous services outstrips supply: A bot looking to transact on a consumer's behalf causes damage to a resource. In this case, demand for pizza during a popular sporting match causes a spike in requests for autonomous vehicle delivery, resulting in missed or late deliveries.

Bots react to mis-processed keywords or reactions and escalate to law enforcement. Bots might be programmed to hunt for certain keywords but react to mis-spelled words and call police on consumers.

Peeping bots will look through keyholes. Unfortunately, custobots will open plenty of vectors for criminals to hack into our home devices.

Scenario 2: Sabotage

We define Machine Customer sabotage as a hard-to-detect code or hardware that can live for a long time in the system before causing a negative action. The most common form is likely to be a hack to a standing order in the Machine Customer's operations, resulting in localized damage or sub-optimal operations that stays under the security systems' radar. Or it can be a series of small actions, often performed by clandestine means, which once they accumulate spur a larger result that would be difficult or unlawful to perform all at once. In this scenario, damage centers on the operation of physical or virtual assets, such as a non-operating garage door or unauthorized communication feeds to damaged industrial equipment.

As the number and reach of Machine Customers increases, we expect their attractiveness as a target will increase as well, which could lead to:

Asset access or tampering: A custobot looking to transact on behalf of a customer (either enterprise or consumer) may open unsecured connections to either the asset or to systems that are connected to the asset that include financial or CRM systems. This may expose the enterprise to issues that range from public relations damage all the way to potential GDPR and sensitive data breach violations.

Security update hacks. Unfortunately, security and software testing often get cut due to budget and schedule pressures on product teams, especially for low-cost consumer electronics. This may enable malicious actors to insert themselves in the loop and take control of the Machine Customer to spy or steal.

Scenario 3: Destruction

In the final (and worst) scenario, the intention is malicious and the damage irreversible. This brings us to major physical damage and/or severe monetary, data or infrastructure loss. To achieve existential threats, damage from second-order consequences is likely to be larger and more dramatic than the first-order damage. Intentional destructive attacks made possible with automated Machine Customers are likely to be done with the motive of eliminating competition, revenge from disgruntled employees or ex-employees, activism and/or political gains. Emotions (such as revenge) will often fuel the motivation for an attack, with power or money as a secondary motivation. Contrary to sabotage, where minimum effort for maximum gain is likely, destruction is more likely to be a sophisticated, targeted attack.

In many cases, the second-order damage will be planned and intentional. Likely damage would be:

- Physical damage to vital infrastructure or machinery
- Major disruption to the supply of vital components or resources
- Denial of Service attacks

- Power outages or manipulation of environmental conditions
- Potential attacks on sovereign infrastructure

Conclusion

Machine Customers will affect markets and their players. In some cases, brand new markets will emerge. In other markets, they may undermine or even destroy existing businesses. With such a revolutionary shift in progress, sitting on the sidelines isn't an option. Like it or not, you will need a Machine Customer strategy.

Key Takeaways

Machine Customers will have an upside.

They could be a colossal upside opportunity for your company. They represent literally millions and millions of new customers ripe for the taking.

Action: Assess the areas where Machine Customers could expand your market opportunity. Will they lead to an expansion of your existing market or open up brand new markets for your organization?

Machine Customers will have a downside.

Machine Customers will also create headaches for your company in terms of technology and processes and spur the resistance of your human employees and customers.

Action: Conduct an impact assessment of Machine Customers entering your market. Will they lead to commoditization or worse?

Machine Customers have a dark side, but there are ways to mitigate risk.

Organizations will need to extend the security measures they have in place to include the dynamics of Machine Customers.

Action: Invest in systems to detect anomalous activity unique to Machine Customers.

Chapter 9

Frenemy 2.0: What Machine Customers Will Mean for You Personally

> Be careful what you wish for, lest it come true!
> – *Aesop's Fables*

Machine Customers will benefit us, but they will also take things away and may affect our careers

Machine Customers Will Help Do Your Chores

How many of us wish we had a personal butler or maid like in the *Downton Abbey* TV show or in the movie *The Remains of the Day*, a trusted servant trained not just to fulfill our needs but to anticipate what we might want next. Landscapers, house cleaners, and babysitters all free us to spend our time in ways we find more fun, fulfilling, or valuable. Over time, we believe the custobots will step in and take care of more and more of the drudgery in our lives.

Table 10 - Examples of Potential Future Machine Customer Drudgery Removal

Note: The machines listed in table 10 do not do the future tasks that we have imagined for them, nor are we aware of specific plans to do so.

Drudgery Task	Example Machine	Potential Future Machine Customer Action
Sweeping and cleaning floors	Robot vacuum cleaner (many makes and models exist)	Reorder cleaning supplies, filters
Washing Windows	Robot Window Cleaner (example: Hobot Window Cleaning robots[1])	Reorder cleaning fluid, batteries, motor repair
Mowing the lawn	Robot Lawn Mower (Many makes and models exist)	Reorder blades, battery, motor repair
Cooking and washing up in the kitchen	Smart Faucet (Example: Kohler Sensate[2])	Reorder water softener cartridges, dishwashing liquid

As the table suggests, we believe all these devices may become Machine Customers. Automated lawn mowers will need new blades, household robot floor cleaners will need cleaning fluids and mop pads, and robot floor sweepers will need filters. These devices will also need battery replacements and repair services at some point. These machines also can collect information about their working environment that can lead to new business opportunities.

Now, not everyone views these tasks as drudgery. We suspect there are people who genuinely like to sweep and clean their floors. They may even get satisfaction from it (Don is not one of them – he has even programmed his Roomba to sweep the floor in the garage). People tend to walk through recurring tasks mindlessly, like zombies, or in a frenzy, as quickly as they can. Don doesn't see doing his laundry as high-quality time, and Mark hates mowing the lawn. If you are one of those people who love these chores, we salute you. But we suspect most people if they had a choice, would prefer to live *Upstairs* than to work *Downstairs*.

Machine Customers will improve your confidence in your buying decisions

When we buy products or services, there is often a nagging voice in the back of our heads that says, "you could have done better." Maybe we could have gotten a lower price or negotiated better terms. Maybe we missed out on something but shouldn't have. How many of us have had that sinking feeling after closing a deal for a new car or making a crucial software investment at work? Marketing professors often call this "buyer's remorse" but at Gartner, we have a term we like better: *purchase regret* – the feeling you have when you ignore your inner voice that says maybe you shouldn't do this, and then you went ahead and bought it anyway.

If you do have that feeling, you're not alone. On the B2C side, buyer's remorse is particularly bad for homebuyers. According to a survey by Trulia, a real estate website, 44% of homeowners felt purchase regret after they bought their home. One of every three home buyers says they wish they had bought a bigger house, while 15% wish they had more information before making the decision.[3] Cars, boats, and extended warranties are not far behind. Purchase regret is not limited to big-ticket items. Even smaller things, like a shirt that doesn't fit quite right or a new perfume that smelled good in the store but not so good when you got home, may be cause for a pang. Business professionals have that sinking feeling too: a recent Gartner survey found that 56% of the technology purchases studied were categorized as high regret.[4]

Why do we experience so much regret? It is mostly because we don't have enough information at the moment of decision. The struggle to organize, analyze and prioritize an overwhelming flood of information while navigating competing options for virtually any purchase leaves customers to struggle with what Barry Schwartz calls the "tyranny of choice".[5] However, it isn't confidence in the brand or supplier they're lacking. It's confidence in themselves, which leads to smaller, safer, less frequent decisions even when buying from a company or brand they already love. In many ways, companies' biggest competitor today is an unconfident customer who is always held back by indecision, analysis paralysis,

or a "good enough" mindset. So how could a Machine Customer, acting on your behalf, minimize purchase regret?
Let's remind you of the key capability of a Machine Customer: their ability to gather and process a large amount of information when they act on your behalf.

Today, search engines like Google, aggregators like Booking.com and cars.com, and e-commerce platforms like Amazon.com and Alibaba.com can present a large amount of information when you research potential purchases. Most will allow you to refine your search and apply filters that narrow the choices – but it stops there. None of these technologies will analyze and trade off a set of options and give you their recommendation.

A Machine Customer, on the other hand, could sort through all the choices and present a set of options - maybe even the best option – for you to consider. You will have the confidence that they did all the research you could not do, and even consider options that might be adjacent to your preferences based on the behavior of others like you, just the way Amazon.com does today: *"People who bought this also bought….* ". But it stops there. What if these companies could recommend products based on what your Machine Customer has learned and knows about your quirks and preferences? Mark's son was shopping for a console for his new flat-screen television. Finding and matching the dimensions of the furniture to the TV required effort since he did not want the footprint of the TV to exceed the dimensions of the TV cabinet. A Machine Customer would routinely know these things and do the research and analysis for us, so we would avoid the wrong purchase decision.

Machine Customers Will Protect You from Your Own Weaknesses

We all do things that, in a perfectly rational world, we should not be doing. Like having that slice of cake when you are on a diet, skipping a workout, or buying that piece of clothing when you have three others just like it. At work, we do similar things. Why do we give in? Some of this is on you, but some of it is the result of carefully developed and tested marketing campaigns. Marketers use researched psychological influencing techniques aimed at your subconscious to direct your choices – or appeal to your weaknesses. "Come on - just one taste" and "solve this business problem with our easy-to-use software!" We all like to think we are smarter than that, but we all succumb to these tactics.

Machine Customers can help you increase your confidence when you buy things and increase your confidence when you *don't* buy things.

At home, a Machine Customer could protect you from yourself by making it easier for you to:

- Stay on your diet by removing temptation from your weekly grocery shopping
- Stick to your budget by reminding you of what you will need to give up in the future if you buy that new purse you saw in the window just now
- Meet your fitness goals by researching a variety of exercise options, including home fitness, gyms, or personal trainers – and giving you the option to select a routine that works for you
- Help you cultivate friend and family relationships by remembering birthdays and important events, recommending gifts or messages, and then sending them

These may seem a bit fantastical, but they are not. As we have said throughout the book, these situations are mere extensions of things we see today – smart recommendations, virtual assistants, and search engines. MyFitnessPal, a mobile app, reminds you to record your meals so you can track your diet.[6] You Need a Budget[7] is a finance application that helps you execute your budget daily and gives you regular progress updates on your financial goals. A Machine Customer will never be able to instill discipline in a human (that must come from within), but they will be able to act as a coach, gently reminding you of your goals when you are about to stray.

Machine Customers Will Take Away Things That We May Miss

Humans and automation have a long history together. When a new technology emerges to make our lives easier, we gain but also lose something. Jobs might be lost, or a personal touch might be missing. Or maybe you like the satisfaction of negotiating the best price for a car or closing an important deal with a client. We will have to pay the same price for Machine Customers as we have for earlier stages of automation: a partial loss of control, choice, and maybe a certain amount of joy.

The Loss of Control

When we delegate a task to someone or something else, we must let go of the direct responsibility for that task and trust that it will happen. Many people are convinced that the way to get something done is to do it yourself. Great leaders and managers have figured out that to delegate effectively and without worry, one must surround themselves with smart, capable, and reliable people.

For example, Gartner's research among serial technology start-up CEOs shows that 35% of tech CEOs (and likely other leaders) feel uncomfortable delegating.[8] We expect some people will feel the same reluctance to delegate tasks to a Machine Customer. In fact, in certain respects, delegating to a machine may be even harder than delegating to people:

- **Retaining "areas of expertise"** — For example, a tech CEO with a flair for sales and marketing may try to justify retaining all marketing activities. This can happen even when the activity has little marginal value to the business.
- "**Only I know the right answer**" — Good decisions add up, but bad decisions multiply. As a result, tech CEOs would often rather stretch themselves than allow others to make a potential bad call.
- **Not valuing diversity** — Some tech CEOs struggle to find someone who shares similar strengths, as they believe that the existence of such a person would make it easier to delegate. Instead, they need colleagues with diverse backgrounds and perspectives.

We expect the same feelings might be true when delegating tasks to a Machine Customer, especially in "retaining expertise" or "only I know the right answer". In fact, delegating to a machine, depending on the complexity of the task, may be even harder than delegating to people. What other consequences might we experience with Machine Customers and a loss of control when we make them our agents?

- **We might not fully understand how the machine makes decisions.** Even if humans set the parameters, we may have trouble explaining why a machine made the decision.
- **We may doubt the ability of the Machine Customer to execute the transaction.** This is about trust in the machine and surrounding technology.
- **We may feel threatened when a Machine Customer could do our jobs better than we could**. Thanks to their ability to process vast amounts of data and weigh it dispassionately, Machine Customers may be able to negotiate better deals than a human.

The Loss of Choice

When it comes to buying things or services, choice is a double-edged sword. Too little choice, and you feel constrained or limited as a buyer. Too much choice, and you feel overwhelmed – a phenomenon that University of Pennsylvania psychologist Barry Schwartz called the "paradox of choice".[9]

How might a Machine Customer limit your choices as a buyer? Well, in some ways, that depends on you as the human set the marching orders of your Machine Customer:

- How much are you willing to pay?
- How many ratings do you need before you feel confident?
- Are there any social, environmental, or political considerations that might go into your purchase?
- When do you need it?

Just like online search, the narrower the parameters, the fewer the choices that come back. This may make you calmer, but there are reasons you may not be able to give up that nagging feeling of purchase regret entirely:

> **A Machine Customer may favor one supplier to the exclusion of others.** If a supplier consistently meets the requirements of the Machine Customer, it may have no incentive to explore equivalent alternatives.
>
> **A Machine Customer may fail to discover new brands or businesses.** Competition is the lifeblood of a market economy, but a Machine Customer might hamper it by filtering out new businesses and even new approaches to business.

As a seller, the Machine Customer era may also be challenging. While the reduced choice may be good for human customers, new brands or businesses will find themselves challenged to be considered by Machine Customers that default to proven suppliers.

Loss Of the Joy of Spontaneity

Shopping and buying things for yourself and others is a source of joy for many people. Finding the perfect item. Landing a better deal than your next-door neighbor. Shopping can be a treasure hunt – a process of discovery.

Neuroscience research tells us that shopping makes you feel good. A growing body of brain research shows that shopping activates key areas of the brain, boosting our mood and making us feel better - at least for a little while. For example, much of the joy of holiday shopping can be traced to dopamine, the brain chemical associated with feelings of pleasure and satisfaction. It's released when we experience something new, exciting, or challenging – and for many people, shopping offers all those things. "You're seeing things you haven't seen; you're trying on clothes you haven't tried on before," explained Gregory Berns,[10] an Emory University neuroscientist and author of *Satisfaction: The Science of Finding True Fulfillment.*[11]

In the Machine Customer era, will you miss finding that special item that you just stumbled upon, or picking up that new candy bar at the grocery checkout stand? If the dopamine hit you get from shopping is lost, will you need to replace it with something else? Shopping has also always been a social activity. In our personal lives, we often go to a mall, a car dealership, or an open-air farmer's market not only to buy things but to be near other people. Once a Machine Customer does the shopping for you, you will have one less occasion to interact with other people. Last, shopping is about giving and caring for others - taking the time to select a bouquet of flowers for a sick relative. Bringing a ready-made meal to your elderly parents. Buying a small present for your child when they have had an accident. Birthday and Christmas presents. There is a joy in the shopping, selection, wrapping, and giving of presents that is very much a part of who we are as humans.

That said, there may be less important people in your life for whom issuing a command to Alexa to send a gift from you will feel perfectly fine. Maybe this will free you up to spend more time with the people you care about most.

Machine Customers Will Impact Professions and Careers

There are almost no TV repair people anymore. There used to be thousands of them, but now they are gone. This is nothing new. Maybe their parents were elevator operators, and their grandparents' telegraphists. Technology changes have always disrupted employment categories and career pathways. The arrival of Machine Customers will do the same.

The key questions are will it make me redundant? Do I have anything to offer this new world? The answers are 1) maybe and 2) definitely. Tim Quast, President and founder of Modern IR, an analytics firm serving the invest relations industry, told us:

> *"Artificial intelligence is one of the core disruptions [to our profession] because it changes the entire nature of the job from one where we are building relationships to ones where we're data analysts – and we are going to have to get as good at counterintelligence data as the machines are. It's absolutely a transformational event for our profession and we either adapt to that or we become obsolete."*[12]

There is no reason to suppose the arrival of Machine Customers will be a net destroyer of jobs. Compare it to the arrival of the personal computer. Before PCs, companies employed armies of typists to write letters and memos. A corporate HQ typing pool was a whole floor full of people sitting at closely ranked desks who did nothing all day but type messages that had been written in shorthand or recorded on a tape machine. Between 1980 and 2000, millions of those jobs were lost. Nowadays, everyone types their own messages as emails, letters, and IMs. But this loss was more than counterbalanced by the vast industry that sprang up in manufacturing and operating personal computers, from chip designers to application programmers to office network installers. Today there are more web designers in the U.S. than farmers.

In a similar way, the arrival of Machine Customers will create many new opportunities even as it eliminates others. The amount of human work that will be required to build a Machine Customer world is simply vast, and they won't all be computer programming jobs. Machine Customers must operate in a world designed for humans. The virtual ones will be 'fed' with images, words, parameters, and ideas by knowledge workers. The physical ones will need to be manufactured, distributed, maintained, upgraded, and recycled.

Machine Customers won't be able to design themselves or deploy themselves. The self-replicating machine is still science fiction. Every context in which Machine Customers operate will be humanly conceived and enacted. Think of all the HP instant ink printers or Illy coffee makers – all the models and designs, all the companies involved in manufacturing them, and all the systems and data people involved in developing and maintaining their connected support services. Imagine how much complex human design and delivery work has already gone into the connected vending machines, AI stock trading algorithms, and early self-driving cars. Yet these are probably only the amoeba of the Machine Customer ecology yet to be built.

Machine Customers will evolve only gradually. Initially, they will often look like smart digital sales channels. The capabilities and competencies needed to bring them to 'life' will emerge over time. This gives us all a few years to adapt our skillset to serve this new world of Machine Customers - but we must be alert to the changes and willing to move with the times. Holding on too long or living in denial won't serve you well.

If you design for, manage, or directly interact with human customers, then you have one question to ask yourself. When will Machine Customers enter my sector, and how will that change the demand for what we do? Directly responding to Machine Customers is less likely to be done by humans, but it will happen. Remember in Chapter 4, the example of the Google demo that had a machine-synthesized voice calling a restaurant to book a table with a human? If the customer you directly serve today can flip from being a human to being a machine, demand for your role may decline. Imagine, for example, that you are a call center expert in a sector where the customers become machines, and those machines start ordering by API. Under those conditions, demand for call center leaders might decline, and you may wish to get ahead of that shift by finding another career path.

If what you do for customers is indirect - such as setting prices, designing packaging, or developing products – then what you do will change rather than disappear. In some markets, complexity may rise because Machine Customers will drive the sector towards more variety and choice or a different pace of innovation. In other markets, the advent of Machine Customers may have a consolidating and standardizing effect. The question you should ask yourself is different: is your market over-extended and made artificially complex to take advantage of the weaknesses and foibles of humans today, or is it inefficient and smaller than it ought to be because of those humans? These broad trajectories can help you frame your future and foresee whether demand for your career will grow or decline. The nature of your work will also change – you may need to consider machine needs more and human needs less. It could be that you will need to pick up training and experiences to help with that transition.

There Will Be Many Roles to Play in The Machine Customer Transition

There will be a need for change leaders of all kinds and at all levels. There will also be new business opportunities for entrepreneurs and new specialties. This enormous transition will require management and commercial leadership at every level of the organization, from executives dealing with operating and marketing changes to engineers who will design generation after generation of ever more sophisticated robot customers.

Business transformations at scale consume hundreds of thousands of graduate-level careers. Consider, for instance, how many people have been involved in helping companies improve their diversity, equity and inclusion behavior. There are senior management workshop leaders and consultants creating change program plans, and sensitivity trainers. Someone had to write the new terminology guidelines and design the new hiring processes. That is just one sphere - environmental sustainability transformations and major banking regulation changes are just as extensive. Each time the government and business worlds need to make a big transition, huge amounts of high-grade work opportunities are generated. Often, the cross-industry transition period lasts ten to twenty years, so an individual who gets involved in the change wave can 'surf' from situation to situation over a long period.

The arrival of Machine Customers will affect different industries at different times, but we think the change wave will take at least twenty years, probably more, given it's bound to be such an enormous technical and cultural shift. For those who want to get involved, there will be career opportunities from now until their financial advice bot starts shopping for their assisted living center.

While we have tended to focus on the impact on business careers so far, we should say there will also be a significant role for government policymakers and regulatory experts to play. The Machine Customer market evolution won't be left to so-called 'free markets" alone. That never happens in practice. From the evolution of mobile telephony to the advent of unmanned aerial drones, governments have played a crucial role in setting rules, guidelines, and standards. Without them, market evolution can take harmful directions, such as control by an exploitative monopoly, or just massively underperform because of a lack of interoperable systems. So, the advent of Machine Customers will inevitably create many roles for specialists who can create standards that will help a stable, competitive Machine Customer ecosystem grow. There will also be roles for people who police those standards and investigate breaches or unexpected situations. Regulators within sectors, such as agriculture and electrical safety, will also need to weigh in on the use of Machine Customers in their domains.

Certain categories of Machine Customers will eventually require whole new job categories. For example, robot wranglers may be needed to find and unblock wayward Machine Customers that have become confused or trapped. That might sound like pure sci-fi, but in fact it's a role that already exists today for Google's experimental autonomous taxis. The Waymo taxi service has been operating on an autonomous basis in Phoenix, Arizona, since 2020. People can call up a ride from their cell phone using the app, and a driverless car arrives at their location. The passengers get in, and the car drives the public streets all on its own to the destination. Very occasionally, as you might expect, things go wrong. Waymo cars are programmed to prefer to stop safely, rather than take a risk to pull out or execute a dangerous maneuver, so they occasionally freeze. When that happens, a specialist from a full-time recovery and safety service drives out to rescue the wayward robot vehicle from its predicament. (A YouTube video from 2021 shows this in action.[13])

It's early days for this, of course, and it will only get more complex. What happens when most cars are autonomous? What happens when they start earning their own fares and spending some of that on fuel, recharge, car washes, and windshield wiper fluid top-ups? An army of specialists will be needed to help these simple custobots when an order somehow goes unfilled.

The Value Focus of Creativity Will Shift Away from Marketing to Humans

If more customers are machines and fewer are humans, psychological appeals to human emotional responses won't be as important a part of marketing as it is now. SEO (search engine optimization) and programmatic advertising specialists will have less of a problem. Their analytical and technical skill sets will be able to adapt easily to Machine Customer needs. However, there could be a problem for those more involved in the traditional arts of marketing.

Right now, while we are writing this book, nobody is worried about that; but they will be. When it starts to happen, it will be a noisy debate. Marketers and PR people have a large measure of control of our societies' narratives. That is inherent to the roles they occupy. When their own roles are changing or under pressure, as it has been, for example, as media has digitalized - they tend to make the news center on their own situation. That's why we expect this issue to become a big story in the mid to late 2020s.

Ever since the 'mad men' of the 1960s, many creative and artistic people have been drawn towards marketing as a place to practice their unique talents. What happens to all that human creative capacity if the demand for it declines? If the target customer is a machine, we will not need clever advertising slogans, cute brand mascots, or artfully lit food photography. There is certainly no need to create a panic about this. With respect to AI, there have been too many 'Chicken Little' narrative books suggesting that the Sky is falling and will land on all our jobs. However, there could be a long-term reduction in demand for this kind of work. That is a policy issue for government, industry, and academic leaders to consider as they develop the next generation of talent. It is also something that should give an early or mid-career specialist pause for thought.

In the United States in 2020, there were 800,000 people employed as advertising, marketing, promotions, PR, and sales managers, including 250,000 graphic designers, 140k merchandise displayers and window trimmers, 120,000K telemarketers, and 270,000 PR specialists. [14] Many of these people focused on the work of psychologically influencing humans to buy things. The Bureau of Labor Statistics forecasts more people will be employed in all these areas in 2030, but we think the arrival of Machine Customers could reduce a lot of that demand by that time.

However, human needs are infinite. We are not suggesting that all these creative professionals will be out of work. Instead, their capacity will be diverted to something new. The downsizing of a marketing agency won't lead to the kind of social problem we have seen in the past with the end of mass employment in other industries as they automated, such as mining or agriculture. We believe more of these people will be put to work designing the product rather than marketing it. Perhaps society will find a way to redeploy some of the human persuasion specialists to focus on changing attitudes in problem areas such as obesity, climate change, and social care.

Does The Machine Customer World Create Opportunity for Me?

That's the big question, right? That is what you really need to know about all of this. The answer could be yes or no. Whatever your current role, the answer is yes – but only if you are ready for what is coming and open to change. Fortunately, the custobot world will not pop up overnight. There will be plenty of time to prepare and adjust.

Do Not Leave It Too Late or Resist the Change

You might tell yourself that you love the human aspects of customer-related work and that machines just are not the same. That is a natural reaction, and you're not wrong – but it could cost you.

Instead, you might ask yourself how much real customer contact there is in your job today. A lot of it is an illusion. The truth is, we already work in arm's length digital and remote e-commerce businesses, often with little direct customer contact. Review how much of your day is really spent in human customer interaction – it may not be as much as you think. Then consider the mode of that interaction. If it is by email or chat, ask yourself, how do you know for sure that the last question did not come from an AI bot anyway?

Sometimes it pays to look at a new situation from a different angle to realize that it can be enjoyable. For example – do you like video gaming? If so – all those game characters you have been interacting with for years are like the AI customers that you will be encountering in the future. Perhaps what you really enjoy about the commercial role you do today is winning in the battle for customers rather than human interaction itself.

The arrival of Machine Customers into individual markets will tend to be non-linear, rapid, and unforgiving. Be prepared to be agile. One of the beneficial outcomes of the COVID pandemic has been to untether work from the workplace a lot more. Do not resist that – take advantage of it. If you make remote or hybrid work your go-to style, it will position you to take advantage of more widely dispersed job 'locations'.

What if you love what you do, it is fundamentally all about human customers, and there is no getting away from that? Holdouts will be able to find work supporting the dwindling number of human customers in some markets for a long time. But you had better make sure you are extremely good because the competition will be keep getting stiffer. As job roles dry up, it becomes a buyers' market where premium salaries are hard to win. The thing to do in this situation is to aim at being in the top 5% of your field. In the past, being an 'upper quartile' performer might have been good enough to keep you safe and happy. Now you will need to think about investing time and effort to become one of the absolute best. Alternatively, you might consider moving to a different country. Machine Customers will catch on sooner in advanced economies. However, in simpler emerging economies, the need to serve human customers may persist for much longer.

It Is Possible Machine Customers Will Cut Your Career Track Short

In the 1930s and 1940s, many people were involved in servicing steam locomotive railways. During the 1950s and 1960s, some of them moved over into modern diesel railways – but the heyday of rail had passed, and diesel trains needed less constant attention, so there were fewer jobs. What happened to all the displaced specialists? Most of them ended up employed in the fast-growing mass automobile sector. So, if you are the modern equivalent of a steam engine maintenance engineer, you must try to accept that your role is going to become obsolete, think more broadly about what your career is, and reset.

Today you do some specialized kind of work for human customers. Tomorrow there will be a fast-growing new consumer segment that happens to be machines. Like the pivot from railways to automobiles, you will just need to find a way to use your talents in that new domain. Lubricating pistons were common to both transport modes. Look for that kind of baseline commonality to reinvent what you do. Machine Customers are new forms of buyers, just like cars were new forms of transportation. Pricing strategies for Machine Customers may be different from pricing for human customers. But it is still pricing.

Conclusion

There is no question that Machine Customers will have a mixed impact on people. There will be positives and negatives to consider, and it is hard to say what the final balance will be. We believe it will be a bit like social media which, over the past decade, has infiltrated most aspects of human life. It's benefits for human connection have been well documented - but its dark side is also very obvious. Like social media, Machine Customers will represent a significant and inevitable change to our way of life. Will the positives of Machine Customers outweigh the risks? We think so, but powerful technology always presents powerful challenges.

Key Takeaways

Machine Customers will make your life better.
They will remove drudgery from our lives, increase our confidence in what we buy, and protect us from our worst weaknesses.

Action: Identify the drudgery in your customers' lives and what saps their confidence. How can the introduction of a Machine Customer make things better?

Machine Customers may also make your life worse.
Machine Customers may also take some of the fun out of the purchasing process. Choice may be unintentionally limited, the joy of discovery lost, and people will be left even more isolated.

Action: Use your Voice of the Customer research to understand the emotions your customers experience when they buy from you. Be mindful that automating some aspects of the purchase may weaken your customer loyalty.

Machine Customers will change our professions and careers.
If you design for, manage, or directly interact with human customers in your career today, keep an eye out for Machine Customers, who may very well be taking part or all of your job.

Action: Consider your current career track. How could Machine Customers accelerate or derail your career plans?

Chapter 10

The Lasting Impact of Machine Customers On the Economy and Business

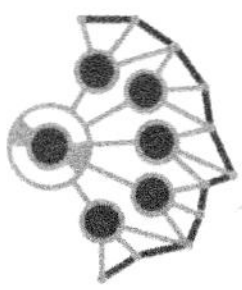

Custobots have needs.
Needs create markets.
Markets make economies.

What will the custobots' impact on the broader economy be? How will they change the nature of business?

The Economic Upside of Machine Customers

Industries and whole economies will be fundamentally more efficient, productive, and larger when Machine Customers get involved. Growth is the prize.

Machine Customers Will Spur Economic Growth Where They Are Embraced

Often in this book, we have written about existing kinds of machines that will gradually transform in the future to become customers. But some of the Machine Customer markets that become apparent early on will be new kinds of businesses that have not existed before. As new forms of intelligent digital products emerge in economies, they will have Machine Customers integrated into their business model from the get-go. Think, for example, about the imminent arrival of electric vertical take-off and landing aircraft. EVTOLs look like the hobby video camera drones that we all know, but supersized, big enough to carry passengers and often intended to serve as short-range air taxis (Figure 13). As futuristic as this might sound, it is also a serious emerging industry. Aviation magazine lists at least ten makers with substantial financial backing, including Boeing and Airbus.[1]

Figure 13 - An Early Model EVTOL

Image: Matti Blume, Volocopter, IAA 2017, Frankfurt, CC BY-SA 4.0 via Wikipedia

From the start, these aircraft are being designed to operate semi or fully autonomously. You will get in as a passenger, and it will fly you automatically, without a pilot, from A to B. Now notice all those fan blades, including redundant ones for safety, that will wear out over time – as all fan blades do. Will humans ever be designated as the primary purchaser for EVTOL replacement fan blades? That seems unlikely to us, not in a world where Tesla cars already pre-order their own parts. It is far easier to build diagnostics into these machines that can sense when replacement is needed, rather than to rely on a human operator to remember to keep checking.

Many IoT-connected services will operate this way. Industrial window cleaning robots will need consumables and insurance. If you can create a machine that is able scale and clean a skyscraper automatically, why would you make it dependent on humans to do something as simple as checking if it is insured to operate or that it has sufficient cleaning fluid? Clearly, as new kinds of digital products arrive, we should assume they will be capable of acting as customers and will be markets in their own right. There will be not only primary economic growth around their creation and deployment but also important secondary economic growth arising from their actions as Machine Customers.

Many new adjacencies and product opportunities will arise because of the existence of Machine Customers. Again, this will lead to growth, as Machine Customers discover new opportunities. Our human economic system relies on people to do something about the gaps, but there's a problem: humans are remarkably good at ignoring facts and data, often preferring to rely on intuition and experience. But that human irrationality leaves gaps – often for an exceedingly long time. We have only to look at the experience of the COVID pandemic and remote work for an example of that. For over a decade, Skype and similar web-based video conferencing systems made remote work possible, but businesses almost completely ignored the opportunity. It took a crisis for us humans to collectively 'find' a pathway to hybrid working and become mass customers of Zoom and other video services. Machine Customers operating in new markets will uncover these kinds of market blind spots far sooner.

In many industry sectors, the arrival of Machine Customers could lead to higher wealth generation through the reduction of costs and higher average gross profit margins. These improvements would come in several ways:

- Lower cost of sale (reduced human marketing and selling overhead)
- Reduced need for human after-sales support
- More rational behavior transaction efficiencies such as optimal economic order quantity and timing
- Better revenue optimization through pricing models that follow more logical and predictable price elasticity curves

The pricing issue is a good illustration of the advantages of custobots. Human irrationality causes complications and distortions. For example, we misperceive the volume of certain pack sizes, we have a sticky relationship with round number prices such as $10, and we get overwhelmed by many price points and options. The relative greater rationality of Machine Customers could enable revenue models that more precisely fit demand patterns.

Machine Customers Cut Waste, Leading to Higher Economic Efficiency

Not everyone will welcome it, but a Machine Customer economy will be more traceable and taxable. "Who" bought what, when, where, and why will be recorded by Machine Customers. In that sense, Machine Customers might be a boon to governments, especially in countries with a large informal economy.

Automatically recorded transactions are easier to arbitrate, leaving less need for slack and less room for fraud in the gray areas. The benefits of technological recording in markets can be very large. Visa Inc measured an 87% reduction in fraud when the credit card industry switched over from magnetic stripes to EMV (chip) credit cards.[2] We tend to trust human economic actors, perhaps too readily. However, we won't trust Machine Customers just because they are machines. Instead, we will design safeguards from the start. The checks and balances of blockchain architectures have already demonstrated ways we can do this. Mechanisms like proof of work and proof of stake are designed in, making almost perfect persistent transaction recording and incontrovertibility key features of the way an exchange takes place.

Human customer markets are often less efficient and smaller than they could be because of human labor overheads. As we have said earlier, one of the key points we want to impress on you is to think of being a customer as work. Shopping is often an endless chore list. Not only is much of it unenjoyable - it also takes up a lot of time. Your reflex might be to think of spending the time that Machine Customers will save you on leisure and consumption – such as TV or golf. However, it is also likely you would use some of the time in ways that contribute to the economy, such as by spending a little more time working or volunteering, perhaps. Throughout history, when any kind of work is automated – it has released talent and labor to move on to other things, and that mechanism has been the key to long-term GDP growth. This will be true of Machine Customers as well.

Machine Customers could reduce buying cycle times, remove unnecessary repeated interaction steps, failed sales engagements, and many other sources of inefficiency. It could also enable faster company formation through lower barriers to entry and the possibility of testing start-up ideas in real markets while burning through lower amounts of venture capital.

Do you look forward to the next in the endless stream of customer rating requests and "Tell us how we did" emails and pop-up messages? No, nobody does. Customer feedback is a chore. Human customers tire and get bored quickly, but machines won't. As the Machine Customer economy evolves, it may be possible for feedback to run to many questions and factors – not just one or two. It's also plausible that Machine Customers would reply far more often, leading to deeper and more complete datasets for analysis. Improved feedback cycles will lead to better performance gap analysis and product variation discovery.

More and better feedback could lead to an explosion of incremental innovations. As more market gaps (prices, product variants, geographies) are identified by Machine Customers, entrepreneurs will find many new opportunities. New company formation – a vital aspect of a thriving economy – could benefit. The advent of human e-commerce created a lot of small business opportunities through Etsy, Amazon Marketplace, eBay, Shopify, and others. Machine Customer feedback could generate new waves of growth through a different style of niche opportunity.

Human behavior is varied and often unpredictable. One of the most popular business leader tools for dealing with such heterogeneity is customer segmentation. This simple, and widely practiced approach breaks customers into groups based on age, geography, or other demographic or firmographic factors, then treats each group as if it had an average coherent set of needs and behaviors. The trouble is that segmentation is often quite crude, and individual customer businesses or consumers do not conform to the stereotypical behavior that has been laid out. Simply put, humans are awkwardly diverse.

This won't be the case for Machine Customers, at least for replenishment orders. For example, each individual HP printer, Tesla Car and Amazon Alexa is very similar to all the others. There are some differences due to tailoring, but overall, these Machine Customers look, feel and behave very similarly from the outset. Business leaders are not faced with the complexity of wide diversity and won't need to do much segmentation and averaging. Even when they do need to segment a Machine Customer market, each group will have well-known and defined behavior. Perhaps a Tesla Model Y will behave differently from a Model 3, but Model Ys will behave consistently.

The rational and reliable operation of Machine Customers is often, though not always, likely to be more predictable. There will be less noise in Machine Customer market data, fewer outlier behaviors and strange anomalies, and better original data quality. All of these will help build better prediction models. That, in turn, will enable better production and supply chain efficiency overall.

As Machine Customers proliferate, the effect should be to add predictability and coherence. That should lead to reduced or more manageable volatility.

Overall, the efficiency of markets is improved, and that should strengthen economies.

On the other hand, too much homogeneity of Machine Customers could be a problem. We won't want them to all do the same thing at the same time. For example, imagine a city in which autonomous taxis operated by Google Waymo and Baidu Apollo roam the streets looking for business. One service station decides to set a price two cents lower for EV power charging than others for a while to drum up business. We can't have every car autonomously converge, causing gridlock. So, there will have to be some collaborative, interactive behavior or deliberate randomness incorporated into the algorithms of Machine Customers. Cars could exchange information so that the close ones notify the far ones that the lines are too long, and the savings are no longer worth the lost time. Or they could randomly delay the time before they set out for the next service station. Or the station could work out an ultra-precise pre-scheduling system, air traffic control writ small. Such programmed random behavior might seem to be emulating a human customer population, but it will be designed-in and cleaner. Distributions will be normal for example, and not skewed. Truly aberrant exceptions to expected patterns will be fewer or non-existent.

Machine Customers will usually complete steps in a process sequence and do so in the correct order. This is not something human customers are good at. How many times have you quietly tutted when the person ahead of you at the grocery checkout ran back into the aisles to get the one thing they forgot? So much of the inefficiency of human markets comes from unpredictable and incomplete behaviors in sales cycles. If there are five steps in completing an online registration – how many humans get bored and give up at Step 4 to go for a cup of coffee and never come back to complete? E-commerce shopping sites across the world are awash with abandoned shopping baskets. Aggregated across economies – the amount of compensating effort used to overcome human customer inefficiencies is huge. For example, returns processing can eat deeply into profit margins, as much as 10% of human customer sales are returned.[3] However, the logical, methodical, undistracted efficiency of Machine Customers should help to reduce that systemic waste. Machine Customers will pay attention; they won't click on the wrong item or order two of something accidentally. They can also help cancel subscriptions for you. For example, companies such as Rocket Money and Experian identify your subscriptions to help you to stop paying for things you no longer need. Once they are identified, it helps you cancel the ones you no longer want, using automation to jump through the hoops that companies put up to discourage you from canceling a service.[4]

Machine Customer-provided data might end up being better in quality and fidelity. Take the simple act of ordering furniture for a room. How good is the average human customer at calculating whether that sofa they see on the page will fit into their front room? Initially, it looks easy – but can it fit through the doorway? Will it make it around the turn at the top of the staircase? Many people have found themselves at one time or another in their lives maniacally shouting "pivot! pivot!" like Ross in that famous scene from the 90's TV sitcom Friends.

Acting as a collaborative Machine Customer system, a smart home could help avoid all the stress and the wasted trips of thousands of sofas being returned to furniture showrooms because they wouldn't fit. Technically, we could already do it today: Common robot vacuum cleaners already use their sensors to build accurate maps of interior home layouts.[5] Amazon Ring is marketing an "Always Home Cam"[6] - an indoor flying security camera drone. Future extensions of this technology could generate 3D spatial maps of home interiors. A Machine Customer would be able to provide the exact measurements to the furniture store, not "oh, that corner is about yay-wide."

Machine Customers Will Create New Business Model Space That Can Contribute to Economic Growth

The internet brought with it some brand-new management ideas and terms, including "business model," "B2C", and "B2B". B2C and B2B were the main business model categories explored in the first phase of commercial internet exploitation circa 1994 to 2004. Over the next decade, two additional categories were added to the lexicon. The first was a new way to transfer value from individuals to enterprises, such as Wikipedia (crowdsourcing) and Kickstarter (crowdfunding) - we might call that space C2B (consumer to business). The second was the enablement of value movement from consumer to consumer — in social networks such as Facebook, peer-to-peer models such as BitTorrent, and sharing economy models such as Airbnb - "C2C". This situation can be graphically depicted as the full elaboration of all combinations of the economic-agent network endpoints B and C (see Figure 14)

Figure 14 - The Four Possible Categories of Conventional Business Model Space

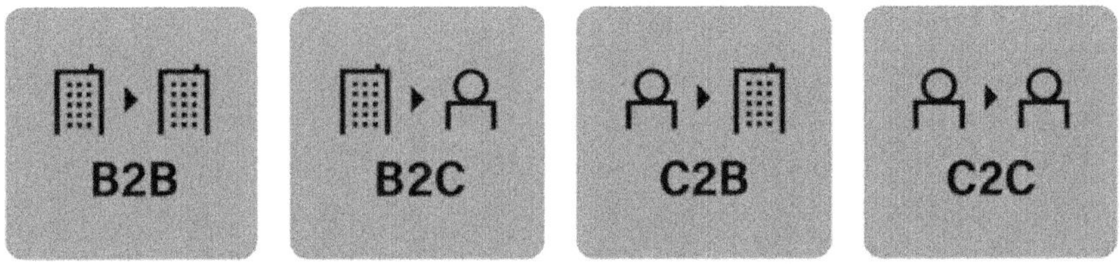

If consumers and businesses were to be the only kinds of economic agents connected by the internet for value exchange, business model category exploration would slow down now. All permutations have been explored — as well as slightly more elaborate business model category combinations such as B2B2C (the description sometimes given to online activities of consumer-packaged goods [CPG] companies).

But as we have identified in this book, AI agents can become virtual customers, and thanks to the Internet of Things, connected products can become physical customers. More generally, they will evolve to become independent economic actors. Let's call them A and T. And that leads to an explosion of possibilities.

Figure 15 - 16 Categories of IoT & AI Assistant Extended Business Model Space

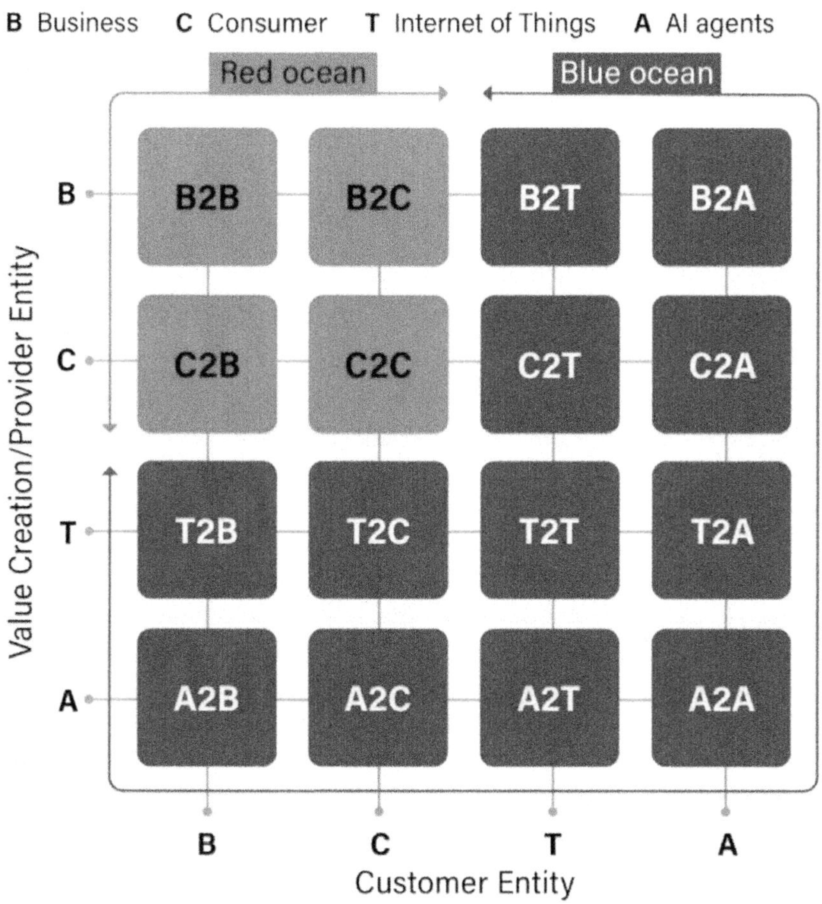

In Figure 15 we show what happens when all the possible permutations of B, C, T, and A are laid out.

Adding Machine Customers Quadruples the Business Model Space

Adding these new actors gives us three times as many varieties of business relationships as before.

Each additional square can be as valuable as B2C (consumer e-commerce and e-service) or C2B (crowdsourcing and crowdfunding) in the long term. This is because there will be many billions of active machines acting as automated or semiautonomous economic agents. These 12 new squares offer a vast amount of what W. Chan Kim and Renée Mauborgne have called "blue ocean" for entrepreneurs to explore.[7]

Consider this: the total economic contribution of the "internet sector" in the U.S. alone was estimated to be over two trillion dollars in 2019[8]. Nearly all of that derived from a 20-year exploration of those first four squares in the top left of the chart - the now maturing and well contested 'red ocean' space. What might the new digital business model category domains look like? Peering into the future can only be speculative, but we can sketch out plausible future business scenarios. It seems most likely the space opened by the growth of the Internet of Things will come first, and the A's will evolve later. For that reason, let's examine the possibilities in two waves.

In Figure 15, we show categories in which things could offer value to businesses and consumers, but also ways in which things could start offering value to each other. The success of these scenarios would depend on platforms, APIs, and messages flowing between B, C, and the machine T and A entities. The machine entities would need a certain amount of algorithmic capability within themselves — or be directly aligned to them in the cloud. The machines would also need a certain amount of defined authority to act — for example, accepting a bid for work, offering data for sale, or ordering a spare part for maintenance. In some cases, the value exchanges would involve a chain of entities. For example, a damaged smart appliance might offer replacement part work via some kind of platform, to which smart 3D printer machines that know their own capability and current capacity might bid. But once a part was contracted and created, the 3D printer might also have to request an autonomous vehicle or drone-based logistics service to make the delivery.

Here are some examples of the kinds of things we might see appearing in these new business model spaces.

A2C — An AI advisor program might become sophisticated enough to act as a valuable bereavement counselor to a human being. This is not so surprising. A combination of Joseph Wiezenbaum's famous 1960s ELIZA experimental program[9] and a modern Siri or Alexa-grade voice recognition and synthesis system could probably create a very simple "sympathetic ear" even today. In the future, an advanced version of a counseling program would be owned by somebody or a company. But it could be granted limited autonomy to seek out its own clients, their approval and consent, and, after some diagnostic matching, offer its services to them.

B2A — An accounting services firm whose partners are regularly helping clients find tax optimization tactics to take the best advantage of the tax codes, steadily builds up a database of case examples. After a few years, this becomes a rich corpus of data it can offer for sale. Neural-network learning systems might seek out, purchase, and use that corpus, blending it with other situational data to train themselves. One such learning subsystem might form part of a "boardbot." This is a kind of AI economic agent first suggested by Gartner analyst Jorge Lopez as a possible future software robot non-executive board director.[10]

T2A — There will be situations in which Ts offer value to As. For example, we can foresee a world in which last-mile drone delivery of small packages and takeout food will become common in urban settings. Imagine police forces augmented by AI software agents. Just as human police officers of old would expect street informants to offer tidbits of intelligence, so the policebot software might expect delivery drones to offer interesting anomalous data seen on their day-to-day travels through their LIDAR-equipped "eyes."

A2A — It's a strange imbalance in today's world that while e-commerce is automating a lot of the consumer endpoint sales work - for example, of check-out clerks and window dressers - there seems to be more work for professional sales executives and purchasing professionals. What if AIs became smart enough to train as proxies for human B2B salespeople and purchasing agents? A great deal of wining, dining and travel expenses might be saved. Negotiation cycles might be reduced from weeks to hours - or nanoseconds.

Use these different scenarios as thought starters in your exploration of Machine Customers and artificial intelligence and their impact on your business model opportunities.

The Economic Downside of Machine Customers

Nothing is all good. There will be winners and losers in the Machine Customer revolution. Even if the overall economic outcome is better, there is always pain in change. Some markets will contract, and there will be transition costs. It is a sad fact that some markets today are artificially bloated by being optimized for human frailties. Especially in the early stages of the Machine Customer revolution, those markets could shrink. Even when we find superior growth opportunities that substitute, the investment cost of reaching those could burden markets for a while.

If we stand back from the amazing complexity, diversity, scale, and success of modern consumerized economies, there is an underlying ugly truth that we often prefer not to acknowledge to ourselves. Whole swathes of our economies are set up to prey on human weaknesses and irrationality and are artificially large. Think about some of the tactics used to sell to us things we may not really need:

- Messages that appeal to primal instincts such as sex, fear, and greed themes in marketing.
- Special offers such as 'buy one get one free.'
- Pricing designed to deceive our value arithmetic reflexes.
- Bright color packaging for children, celebrity endorsement for adults.
- Tiny writing that helps us ignore key information – such as calories.
- Vitamin pills and cosmetics that science says don't work.

Markets that have overgrown to the point of dependency on these kinds of tactics will be precarious in a Machine Customer world. The efficiencies that Machine Customers bring will halve some industries and decimate others. At the extreme, Machine Customers could eliminate some industries entirely – if their whole value proposition is, in fact, bogus.

One example of such an industry at risk is product replacement insurance. Consumers are often sold such cover on electronic devices and white goods at electrical stores, and it is often an irrational human choice market. The known MTBF (Mean Time Between Failures) of the item being bought and the cost of replacing it (probability x price) means it would be far better not to bother insuring. In two years' time, Moore's Law says the next version of the product will be superior for the same money or will cost less. A Machine Customer would be able to 'see' all that very easily. Human customers are too easily lulled into believing that insurance is a smart choice.

The relentless rationality and speed of Machine Customers will have a downsizing impact in many ways. Think for a moment about the call center industry. In the United States, about 2.8 million people worked in call centers in 2020.[11] As more machines do the customer work, what's the need for a human sales or customer service person? Machine Customers will more often solve problems and ambiguities via search, APIs, AI text analysis, and other online computational means. We estimate that by 2027, most sales and service centers will be fielding calls from Machine Customers. This means companies need to set up an easy-to-access online info hub that speaks the same language as machines.[12] This hub should be ready to answer questions coming through things like APIs and data files. Occasionally, perhaps, a voice synthesizing Machine Customer will call a human to ask a question – but that kind of interaction will be an uncommon transitional communication, not a long-term solution. As Machine Customers advance, the need for human call center workers will gradually decrease.

Of course, human capacity is very precious, and it will be reused to better effect in our societies … eventually. But for individuals and some whole communities, the displacement process could be hard. This isn't a new phenomenon. From the 19th century destruction of farm laboring jobs to the late 20th century hollowing out of the Detroit car industry, we have always lived with technological dislocations. But that knowledge won't make it any easier for those affected by Machine Customer changes.

Markets Will Be Commoditized If Their Products Are Artificially Distinguished

In the introduction to a 2014 drinks industry research paper, the authors wrote:

> *"We show that when consumers taste blind, they cannot distinguish between three major competing beer brands. Our results suggest that brand loyalty in this market is likely to be driven largely by marketing and packaging, and not by the underlying sensory properties of the competing products."*[13]

When you browse a shelf for beers in a supermarket aisle – the shapes and colors of the cans and bottles evoke subtle recollections of ad campaigns designed to associate the brand with certain feelings and beliefs. But the flavor profile of cold lager style beers is so similar people can't distinguish them in blind taste tests. So, what will happen when we start delegating beer ordering to a machine? That part of the price due to branding will tend to be reduced over time. The human drinker for whom the beverage that has been purchased, just won't care quite as much as they used to about such things. The machine will know what tastes, quality, and ingredients you like. It won't be so easily manipulated. Of course, brands are important, and people can be stubborn. When they can't be weaned, their preference can be noted by the custobot.

For a long time, the commoditizing effect of Machine Customers will only be partial. It might have a similar level of impact as supermarkets' generic versions of products. Those do not completely supplant major brands, but they do challenge them and tend to reduce margins. Brand marketers will still spend money to influence human consumers, so consumer custobots will factor in what those humans say they want. But in the long term, we think Machine Customers will gradually reshape the relationship between consumer and producer.

We think that brand is still important to a Machine Customer, just in a different way. If we think of a brand as a promise - then can we not make a promise to a Machine Customer just as easily as to a human customer? Perhaps brand marketers will need to create a shadow brand for custobots. One that is based on the promises it must keep with machines – such as data that is accessible, accurate, and timely. This data could also include reading the reviews from human customers and assessing intent. We could imagine a large industrial manufacturer having a brand promise for its human customers based on a set of important attributes. It could also have a variant of that promise articulated for its machine customers. Together the regular brand and the shadow brand create differentiation in the marketplace.

Some markets rely on place and time convenience to win price differences. Indeed, this form of revenue management is a key discipline in some industries, such as hotels and airlines. Machine Customers will be forever watchful and patient. They are equally happy to buy at 02.57 as they are at 13:15, which most humans aren't. Machine Customers can check or compare prices every day or every hour.

This kind of software-based buyer efficiency already exists today in financial markets and parts of eBay. But as it grows, it will affect more sectors. Often industries have come to rely on managing a portfolio mix of customer segments with different needs and price elasticities. Sometimes there are embedded cross-subsidies. For example, a lot of long-haul aviation relies on high-margin business class seat sales to help support lower price fares in economy class; with the total mix creating sufficient demand to fill the plane, making the flight commercially viable.

The endless vigilance of Machine Customers and their ability to seize on price, time, and perishable inventory changes will likely have a pressuring effect on the models of businesses that have previously relied on the relative inattention and inefficiency of human customers.

The ruthless new efficiency effects of Machine Customers (particularly if they sweep into a market quickly) might lead to the failure of individual companies that cannot adapt fast enough. Sometimes there might be real damage to provision from a rapid 'race to the bottom' effect. That could lead to negative economic outcomes. It's easy to say that if a company fails, the process of creative destruction will reallocate its resources – new entrepreneurs will eventually come along and take up the mantle. But when markets are mature and highly concentrated oligopolies, there can be a real pain in the transition. Sometimes the rapid uptake of Machine Customers could lead to one player going out of business in a market where three companies control 80% of the supply. If the goods or services are systemically important, such as fertilizer, baby formula, or natural gas - that could dangerously impact an economy or lives. On occasions, governments might need to step in to slow the transition to Machine Customer control.

Oligopoly Behaviors and Concentrated Risk Might Present Economic Management Challenges

Machine Customer markets might not be easy or cheap to serve. The technology prowess needed, and the capital required to serve these markets, could be high. We have often seen technologically bound markets become duopolies or oligopolies quite quickly. The scale economies needed to make such markets work profitably can be challenging. Take the mobile smartphone operating system market, for example. In the late 2000s, there were multiple players – including Blackberry, Nokia, and Symbian. Today, only two dominate – Apple iOS and Android. Professional personal social networking only has one serious player – LinkedIn. Companies such as Airbnb and Uber do not quite have such uncontested dominance, but the concentration is also high.

Machine Customers must operate on behalf of their users in accordance with well-defined laws, which we will discuss in the next chapter. The creation and maintenance of Machine Customers won't be conducted by those users themselves. You didn't program the Instant Ink service in your printer, and you won't be the creator of the tire buying capability of your autonomous car. The providers of Machine Customer capability will be companies that seek the kind of scale effects we have seen in earlier online platforms. As we have already explained in Chapter 5, lots of independent walled gardens won't win.

Many have argued that the concentration of societal power in a handful of digital giant corporations such as Google, Facebook, Tencent, and Alibaba has gone too far. But in those situations, they have enjoyed power over the supply side through their market brokering. If similar entities were to take control of

Machine Customer provision, they would also control the demand side, and that could be very problematic. It seems likely to us that strong legislation and regulation will be needed from early on. Without it, economies could be stifled by a new variation of quasi-monopoly power that might start with high innovation but, later on throttle it.

When we concentrate activity into machine markets, there is often a risk of unintended runaway swarm-like action when all the machines start to do the same thing at the same time or fall into thrashing loops. As we have discussed, the US stock markets now have 'circuit breakers' designed for safety – but it took a big accident before action was taken. That always seems to be the way with major industrial economic progress – society needs a shocking wake-up call before acting on something that threatens us in its ubiquity – think of DDT, asbestos, leaded petrol, greenhouse gasses, forever chemicals – there have been many examples. We always find societal solutions eventually, but patterns of history suggest there will have to be a Machine Customer market crisis before we react.

Humans are infinitely diverse. That simple fact is innate to biology. Only identical twin humans might be the same by DNA at birth, and even they become rapidly differentiated in their behaviors through their life experiences. By contrast, machines can often be identical or very alike. Designs become staid. Solutions are similar. Parts are sourced from the same place. Every machine runs the same software version with the same algorithms.

The behaviors of human markets can adapt because some humans will always see things differently to the majority. Even if only a few people think differently, they can discover a different insight or pathway that, later, others can follow. This is something that the clothing industry relies on. Human fashionistas take risks, discover designers, and wear things others would not dare. They start the new trends that others eventually follow. But what if humans delegate all that work to machines?

There is a risk that all Machine Customers could suffer from the same monotonic blind spots and destructive repetitive behaviors. Algorithms go around the same search space, never breaking out into genuinely new patterns. Creative new products and alternative solutions to customer needs could be 'locked out' by the repetitive lack of exploration exhibited by Machine Customer markets. And though it is true that neural network-based software systems are capable of machine learning, that usually only happens in the workplace of the data scientists developing those systems. Very few deployments of neural network-based AI carry on learning from the real world as they are used.

Until that AI hurdle is overcome, there will be a downside risk that Machine Customer prevalent markets will behave too monotonically, and that lack of variation might dampen the previous creative dynamism that flowed from human customer diversity.

Conclusion

The arrival of Machine Customers could spur a substantial and sustained period of economic growth. Fundamentally, their arrival will contribute to productivity - the bedrock of GDP growth. The value of that growth is likely to make Machine Customers a subject of international economic competition. However, the value Machine Customers add will only be possible if they are tamed to operate in thoughtfully regulated, free and fair markets. But most of those downsides, if they arrive, will arrive in the distant future. In the short run, for business thinkers and strategists in many industries, the arrival of Machine Customers and new business model spaces will spur a gold rush of creativity.

Key Takeaways

Machine Customers will grow individual businesses and whole economies. The efficiencies and opportunities they bring are too good to be ignored. The only real question will be when - not if - and who gets there first.

Action: business and government leaders should make Machine Customers an explicit agenda topic of conferences, forums, think tanks, and policy-making bodies that create and support industrial strategies.

Machine Customers will open vast new areas of business model possibility. A great deal of innovation and growth was brought about by business and consumer value exchanges over the internet. The addition of two variants of a new kind of economic actor, in physical and virtual forms, potentially quadruples the market space in which businesses can get creative and thrive.

Action: business strategists and CEOs should apply the enlarged map of business model space to their industry, in mid and long-range planning exercises.

Machine Customers will have an economic downside, but it can be mitigated. Machine Customer markets will tend to move quickly towards oligopoly. That power could be abused. Some industries may find the addition of a more logical customer will unpick their prior market optimization around human irrationality.

Action: regulators should be proactive and vigilant in mitigating the commoditizing and market consolidating effects of Machine Customers. Business leaders in consumer industries should embrace sustainability thinking now to gradually wean their business away from reliance on the waste and excess of bloated markets.

Chapter 11

Where Machine Customers Will Lead Us - The Long-Term Future

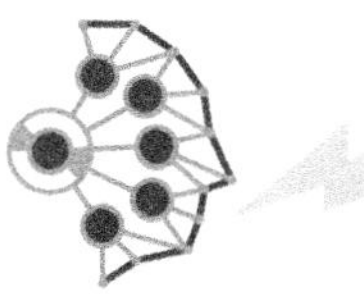

Custobots can help explore brave new markets and go where no human has gone before.

What starts as a cool idea you explain to others, will grow into a trusted aspect of everyday life, and eventually, part of the future of all humanity

This Is Powerful - Pass it on

When any megatrend arrives, organizations are often faced with a lot of confusion. What is this new thing? How is it defined? How is it different? Does it really apply to us? If it does apply to us, how should we respond? What are our options?

When the concept of a Machine Customer starts to become real in your sector, what will most of your peers do? The answer is probably – not much. Most will try to stick to what they know. Early adopters have a clarifier on the executive team– the person who can see the coming changes more clearly than most people in the company and explain them.

Taking on this role could be your first and maybe most important Machine Customer opportunity. When people see that you have some insight, they will tend to empower you to lead them over the first hurdles. Don't be afraid if you can't explain all of it now. Nobody can. But as you start to explain the custobot concept and work with your colleagues to come up with your company's response, you will learn. Practicing different ways of explaining something helps strengthen your own understanding of it, and your confidence in it. One great technique for improving the way you explain is to practice doing so for different ages and levels of experience. For example, if you can explain a subject to a child, an adult, and another expert, you have anchored your own understanding at different levels of abstraction and simplification. That range would then allow you to help almost anyone else.

Try explaining the concept of a Machine Customer to a small child, little Monica, age six.

Monica, you know how sometimes Mommy and Daddy get grumpy when there is no milk in the fridge? One of us has to drive all the way to the shop because we forgot to buy some. And do you know how when you ask her, Alexa – over there in the speaker who you talk to sometimes, and she tells you a joke? Well, now we can ask Alexa to remember the milk and always make sure we have some in the fridge. Isn't that nice?

Next, try explaining the concept of a Machine Customer to Ben - a university student is in his second year of a business degree, and the subject is new to him.

Ben, customers don't have to be people anymore. These days with AI and other technologies, it is possible for a machine to do the work of shopping and buying. Most stocks are already bought and sold by algorithms. One day many things will be. Can you imagine machines acting as customers for consumers and in multiple industries? What implications might there be?

Finally, tell the concept to a chief marketing officer or business strategy officer.

It's often difficult to convince people to buy your product – or I should say, any product. First, you had to convince them that they want something like your product, then that they want your particular product, and then that they need to order it now. The size of your addressable market is fundamentally constrained by that friction. Machine Customers are a way to overcome that. Creating machines that do the shopping and buying for humans does away with that inertia. They will repeat-purchase more reliably and more often, leading to higher, more stable, and predictable revenue flow. But you will need to address Machine Customer markets differently.

Explain Machine Customers Using an Evolution Timeline

People can often understand something more readily if you explain its origins, how it is evolving, and where it is heading. Having that full context helps them anchor the ideas. Here are three different ways to do that.

Digital channels: In the old days, you needed to navigate a website to find what you were looking for. Then they added search functions to find things with keywords. Now they suggest things you will like at the top of the page, or order through their smartphone app. Eventually, you will be able to ask the channel to take over shopping for you altogether – within certain constraints.

Physical machines: It used to be that cars went where you steered them, as you followed the signs or checked a paper map occasionally. Then they started to offer you directions via GPS. Next, they will be able to do some of the driving for you, and they will decide where and when to next fill up or recharge.

AI Agents: First, you'll ask Alexa to buy toilet tissue. After a few times, she will suggest you set up a repeat order for toilet tissue. Eventually, she starts analyzing your shopping list and proposing to take over your other shopping for you. One day, you find yourself asking her why she changed the toilet paper brand, and she will explain that the new brand is made with recycled paper, and she gathers from your other purchases that sustainability is important to you, so she switched.

Next, We Must Trust and Let Go to Grow

When machines take on the work of humans, something is lost in the process. We risk becoming less human. On the other hand, some say that being freed from tedious or repetitive tasks frees humans to pursue higher-value opportunities. Jeff Bezos, Chairman of Amazon, said:

> *"I predict that, because of artificial intelligence and its ability to automate certain tasks that in the past were impossible to automate, not only will we have a much wealthier civilization, but the quality of work will go up very significantly, and a higher fraction of people will have callings and careers relative to today."*[14]

The next challenge will be winning over your human customers to the idea that you don't need them anymore. In the majority of interviews that we conducted for this book, and in discussions with clients and colleagues, the topic of trust was a common thread. Sherry Aaholm of Cummins Inc. told us,

> *"To have an engine just go on its own and schedule a service appointment and show up without any human intervention is not something our customers want at this*

> *current point in time. Technically it can be done, but the level of trust from a customer side is just not there. You might be able to get there over a period of time after demonstrating it successfully."*[15]

What will it take for humans to trust and embrace Machine Customers? Based on our research, it comes down to three things: trusting the technology, trusting the ethics of Machine Customers, and trusting yourself.

Trusting The Technology

For Machine Customers to scale, humans must trust the underlying technology and the contextually aware decisions made by the machines. Humans must trust Machine Customers as much as they trust people who perform services for them. custobots will need to execute human instructions - including implicit instructions - and act in the best interests of humans. For example, when we ask a Machine Customer to order us coffee, we will expect it to include milk and sugar based on past behaviors. We must have confidence that it will do the right thing.

So how do you build trust in Machine Customers? Simply, the technology must work, and the rules governing its use must be clear. For example, we've said that "hallucinations", fabrications, and factual errors are the most pervasive problems with Generative AI. And potentially Agentic AI. Until more automated tools like Guardian Agents are widely available to ensure user trust, organizations will have to manually review model output and mostly use it as a first draft tool. Machine Customers that repeatedly and consistently deliver in all channels, and at every touchpoint, will earn consumer trust. In a B2B setting, trust in a Machine Customer will be based largely on whether it fulfilled your SLA (service level agreement). Yes or no. Perhaps the human that owns the Machine Customer will set parameters for meeting the SLA. Maybe 90% compliance would be good enough. Or maybe it is 100% or nothing. Remember, the Machine Customer will always execute against the needs and objectives of its human owner.

Machines will also need to be programmed to develop trusting relationships with their suppliers, just as humans need to develop a trusting relationship with their Machine Customers. If a supplier fails to meet the expectations of a machine, that machine is likely to make an immediate change, based on the perceived impact to the humans it supports, and do it without emotion.

The human owners (either the end-user or the Machine Customer's beneficiaries) must also trust the custobot to act independently of undisclosed business influences. What if a soft drink manufacturer offers a free repair to a vending machine on its next malfunction, on condition that the vending machine agrees to carry its fledgling ginger ale product for another six months? In this scenario, the machine is no longer working for its human owner, and as a result, the human customer might put less trust in the machine.

To be clear, there will be technology challenges to scaling Machine Customers. One significant challenge lies in talent scarcity, as the sophisticated software engineering skills needed to scale these complex, interconnected systems are difficult to find.[16] Even with the right talent, the necessary platforms for orchestration are largely absent; today's environment favors vendor-specific "walled gardens" when widespread adoption will require neutral market spaces or platforms where machines can transact openly and at scale. Furthermore, testing and validating the unpredictable interactions of numerous agents at scale moves beyond traditional methods, demanding complex simulations to understand system behavior. Businesses must navigate these technology and talent barriers to unlock the full potential of Machine Customers.

Trusting The Ethics of Machine Customers and Privacy

Next, people will need to feel that their Machine Customer is acting ethically. If Machine Customer decisions are based on repeated algorithmic patterns, then businesses, humans, and other machines may have an opportunity to predict their behaviors. That knowledge creates opportunities, both ethical and unethical. In some cases, bad human actors could harness Machine Customers to act badly. For trust in Machine Customers to grow, ethics will need to be hard-wired into their operating system.

Digital ethics comprises the systems of values and moral principles for the conduct of electronic interactions among people, business, and things. Digital ethics is not about the distant future of technology, with dystopian discussions on how robots will take over. Digital ethics is about current practical issues. Ideally, business leaders will identify ethical issues early, avoiding the most difficult dilemmas. Still, unintended consequences and real dilemmas are inevitable. We suggest organizations implement digital ethics by design using the four pillars of "Care Ethics"[17]:

Empathy: How you show you care

Responsibility: How you deal with unintended consequences

Competence: How you take responsibility

Trust: Whether others have confidence in your responsibility

All four pillars are related. Empathy without responsibility is empty. It's like saying, "Oh, I am so sorry", and then moving on. Nothing happens. Conversely, taking responsibility without empathy is mechanistic. You act because you have to, based on regulatory compliance, not because you really want to or really care. Without competence as the third element, empathy and responsibility are useless. You must be able to tackle the matter at hand well. Trust is needed to make the other three work. It is great to take responsibility, but if your stakeholders do not trust you to do so, your offer will not be accepted.

For example: Consider an autonomous vehicle designed by a manufacturer to gather data about repair costs from every repair shop it visits. The manufacturer could then use this information to undercut the local repair shops in the market to favor its own dealerships. Would businesses knowingly engineer this in their machines? The answer is yes, of course, they would. For thousands of years, unscrupulous business owners have demonstrated they will do what it takes to get business. Caveat Emptor will be as relevant for Machine Customers as it is for people.

What happens if ethical boundaries are crossed with Machine Customers? Humans may hesitate to share their data if they think the machines that serve as their proxies are being taken advantage of. In most scenarios, the issue of digital ethics will apply more to the morality of businesses than it does to that of Machine Customers. When customer decisions are completely algorithmic, all those businesses must do is identify sale triggering patterns to win a sale. While there are scenarios where this could work to the benefit of customers as much as to that of the businesses, it also means that Machine Customers are just as susceptible to being taken advantage of as your elderly family members who browse the home shopping network.

Trusting Yourself and What It Means to Be Human

Machines are good. Humans are better. Machine Customers are just the latest in a long list of tools that have made life progressively easier for us as humans. As we mentioned in Chapter 2, the rise of conversational platforms and virtual assistants is the clearest indicator of what will be possible in the world of Machine

Customers. Systems capable of learning more about our purchase and consumption behaviors and recommending the next actions we can take to reduce more of the drudgery in our lives are appealing. If you are like most of us, delegating the purchasing of laundry detergent does not challenge what it means to be human any more than not having to wash our clothes in the river. Human intelligence, creativity, and initiative will always win out over machine intelligence, for the very fact that machines today are not creative, they can't take the initiative unless programmed, and their intelligence has real limits. We should take confidence in this. As Ernst Fischer[18], an Austrian intellectual, critic, poet and political leader, once so eloquently wrote:

"As machines become more and more efficient and perfect, so it will become clear that imperfection is the greatness of man."[19]

We could not agree more. This is why we will not feel threatened by letting the Machine Customer buy our soap.

Entrusting Custobots with Environmental Responsibility

All new technologies and technology-enabled concepts can be used for good or for ill. Historically they have also had unintended, unforeseen, or willfully ignored secondary downside consequences. For example, what impact will Machine Customers have on humanity's desperate need to clean up its environmental act?

Adding complex electronic and digital capabilities to previously inert products will have consequences. On the downside, the Internet of Things expands our use of rare metals, toxic chemicals, plastics, and all the other materials that pile up as vast mountains of e-waste. If physical Machine Customers are to do more good than harm, they will need to be provisioned through a much more sophisticated circular model of production than we have today. Even today's ambient, apparently 'non-physical' AI virtual assistants have consequences. The Cloud that supports their AI will consume a huge amount of electrical energy. As we have seen with the rapid rise of cryptocurrency mining, it is perfectly possible for a computing trend to have large material effects on global energy use. If that energy comes from burning fossil fuels, the Machine Customers will be adding to add to the sum of CO2 air pollution that is driving climate change.

On the other hand, there are many ways in which, if we choose the right path, Machine Customers could help reduce waste and improve our collective environmental footprint. The algorithms within Machine Customers can be tuned to shop more responsibly than we do. They can help reduce the massive waste of consumer overbuying. They can help by calculating and considering complex trade-offs – from food miles to forever chemicals. Is it better to buy soft fruit flown hundreds of miles from a warmer country or grown locally in a heated greenhouse? We don't know, but our custobots might. Machine Customers could be programmed to compensate for our psychological weak points, finite attention spans, and all the other behavioral shortcomings that lead us into environmentally damaging consumption.

We Will Always Need an Off Switch

From the first time we started presenting our ideas about Machine Customers, we have always been clear: humans will always need an off switch. If we cannot stop a machine from executing a task, we put ourselves at the mercy of the machine. The Future of Life Institute agrees. An outreach organization working to ensure that tomorrow's most powerful technologies are beneficial to humanity, the institute is focused on keeping artificial intelligence beneficial for human society while reducing its risks. Best-selling author Max Tegmark, President of the Future of Life Institute, puts it this way:

> "*Everything we love about civilization is a product of intelligence, so amplifying our human intelligence with artificial intelligence has the potential of helping civilization flourish like never before – as long as we manage to keep the technology beneficial.*"[20]

Machine Customers will be important, but humans must always be in charge because trust must be built steadily over a long time. Therefore, an off switch needs to be present in any custobot we make, sell to, or service. Industry standards for safety, liability, and control must be set - whether through self-regulation or government intervention. But we also need to be able to shut them down.

Enter Autonomous Business – Machine Customers Will be Central to this next era of Business

In 2015, Gartner coined the term Autonomous Business. It was part of our preliminary work on our concept of Digital Business. In our roadmap for digital transformation, we knew that there would be a business era marked by AI, but we weren't quite sure what would happen, largely because the technology was not mature enough for scale applications.

Autonomous Business is a strategy that uses self-improving, adaptable technology to make decisions, take action and create new value. This isn't just about adding more automation; it's a fundamental shift towards building upon the digital business era we've seen unfold over the past decade. Technologies like AI, robotics, and deeper automation are forcing businesses to rethink everything from competition, how they create value, to their fundamental operating models. AI is going to be the catalyst for the genesis of Autonomous Business.

Autonomous Business isn't a single technology but a confluence of five key elements These are Autonomous Operations, Augmented Leadership, Auto adapting Products, Machine Customers, and the Programmable Economy. As we've been discussing throughout this book, Machine Customers are a key piece of this future, representing not just what we sell, but who we sell to. This coming era of Autonomous Business isn't some distant sci-fi fantasy; it's already taking hold and will define the competitive landscape over the next couple of decades. It leverages the foundational investments and skills developed during the preceding digital business phase. Gartner's CEO survey indicates that CEOs are acutely aware of this shift, with a significant majority seeing AI as the primary disruptive force in the near future. Autonomous Business represents a significant leap forward, a "step change" and a "force multiplier" that will drastically alter who wins and loses. Organizations that are quick to embrace higher levels of automation and autonomy, particularly by incorporating machine customers and leveraging AI, will gain the advantage in the coming years.

Eventually, We Will Free Machine Customers to Expand Our Horizons

The authors are both unashamed sci-fi lovers. Don's array of Star Trek collectibles is a wonder to behold. The futures imagined by sci-fi authors often illuminate annoyingly plausible technology disruptions that aren't so far off. Peter Schwartz agrees - he told us that many of the seemingly far-fetched technology ideas he wrote into the script for the Tom Cruise blockbuster movie Minority Report (set in 2054) became real by about 2020.[21]

One of our favorite TV series of recent times is *The Expanse*. It is set in a future where humans have explored, settled, and inhabited much of the solar system. Even a decade ago, that seemed like a 1960s retro fantasy. But today, Elon Musk and others are making it seem partly imaginable by the end of the century. That might feel like forever away to you, but it's not too far away for policymakers and long-range planners to start considering. Where will the mines of the future be? What could be manufactured better and more safely away from Earth? How can we generate all the energy we need without frying the earth in a blanket of heat-retaining gases? Exploring and harnessing the great dark expanse seems like a long-term inevitability for humankind, but there's a big problem. Space kills humans.

It costs so much to put a human into space that, so far, only the most powerful governments and billionaire tourists have managed it. Even if Musk, Bezos, and others succeed at making rockets reusable a hundred times, the price to get there will still be beyond the life savings of the 99%. Nobody believes the next price/performance technology step - a space elevator - is less than a century away. But even if we do eventually succeed at escaping earth's gravity for a reasonable fee, it's a ticket to a truly toxic place. Living off earth would mean surviving in an entirely synthetic world. Everything you eat, every breath you take, and every inch of space you inhabit will need to be manufactured and artificially maintained. There is no such thing as a partially natural habitat anywhere. A lack of gravity will wither the human body, and inescapable radiation will corrupt DNA, both dramatically shortening expected lifespan. Because of this, we cannot really conceive of space as a realm where human animals will be the majority inhabitants. Instead, machines will grow the space-based economy for us.

Any economic system requires both users and suppliers, providers and *customers*. It is very common in space-based sci-fi to see depictions of robots as providers of capabilities and services. That part is correct. However, the idea that there will be mass humanity directly consuming in that environment makes no sense. We will never be able to afford to be there except as occasional visitors. There will have to be a mass of machine customers if the 'expanse economy' is to work.

An almost entirely machine-to-machine space economy is not hard to imagine. It is on the cusp of starting in low earth orbit today. Satellites need fixing, upgrading, and deorbiting. A lot of space junk needs to be collected (Figure 16).

Figure 16 - A CGI Image Of All the Space Junk-Collecting Work to Be Done

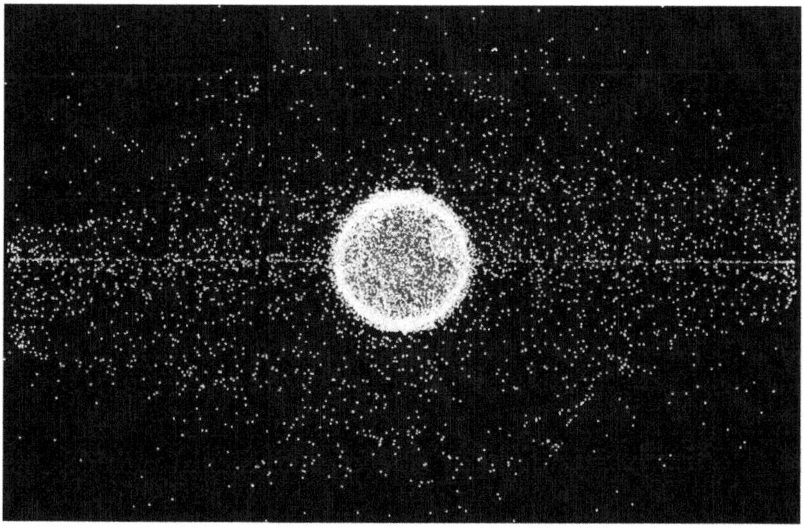

Source: NASA ODPO, orbitaldebris.jsc.nasa.gov.

Companies are setting themselves up to provide services to satellites. Of course, initially, this will all be organized and negotiated down here on earth between humans. But soon enough, that will become a problem. We are too far from the action to manage it well. We will need specialized servicing spacecraft to negotiate, price, transact and serve satellite customers more directly. Add the communication latency of moon orbit, and that requirement becomes more important. Add the latency of Mars orbit, and a directly trading machine-to-machine economy becomes the only way a Martian economy could realistically operate at a cost that would scale.

So, Machine Customers will be a critical component that enables our economy to continue growing off-earth. What is that realm worth? Morgan Stanley estimated in 2022 that the space economy will be worth around a trillion dollars a year by 2040[22]. However, even that could be quite conservative because it's built on extrapolating current satellite uses of space. Machine Customers could also help enable the start of extraterrestrial mining, construction, and manufacturing. In the long-term, such a scenario is not outlandish. Jeff Bezos, probably the most successful entrepreneur of the century, sees it as a realistic future. As he said in 2016:

> *"Energy is limited here. In at least a few hundred years ... all of our heavy industry will be moved off-planet," Bezos added. "Earth will be zoned residential and light industrial. You shouldn't be doing heavy energy on earth. We can build gigantic chip factories in space."*[23]

Long before all that, Machine Customers will also help grow markets here on earth by becoming free roving semi-autonomous economic actors making their own complete contribution - performing as both earners and spenders. We humans are choosing to reduce our birth rates to below replacement, and at the same time, we want the lifestyle improvement that ideas such as a four-day work week might bring. But we don't want to sacrifice advancement in living standards. To meet this societal challenge, Machine Customers will do work that we need done but no longer wish to perform, such as cleaning. Think of them a little like the way sheep manage grass landscapes for us, or bees provide 'pollination services' for farmers. Here's a simple 'news from the future' fictional scenario for the humble street trash bin.

5 May 2037 PRESS RELEASE:

Today, 5000 LiveBins were released on the streets of Chicago.

The release of LiveBins was hailed by the mayor as "ushering in a new era of low-cost cleanliness for the city". The 5000 LiveBins were 'freed' (deployed) over 24 hours at random spots around the city.

They will each be paid in crypto for the amount of trash they 'eat' [collect]. Bins will call up and pay for autonomous trash trucks to empty them when full. Bins will migrate themselves to where the hottest 'trash action' is in the city by collaboratively sharing demand data so that they exhibit flock-like behavior. For example, more bins may move to the parks in summer. They will each move independently by requesting an autonomous van pickup and drop-off and paying for it with the crypto they have earned. They will negotiate for sidewalk space and pay rent [competing with other kinds of future living street furniture such as benches and outdoor ad displays].

The LiveBins are expected to live on the streets this way for an average of twenty years. They will also request and pay for their own 'health' services, such as repairs to damaged parts, to maintain their longevity. Providers of these services will have complete access to the LiveBins fully open-source design. LiveBins will maintain a minimum crypto balance sufficient to order their own 'funeral' and 'rebirth' (end-of-life recycling and replacement).

While LiveBin Inc. will remain the nominal designated owner and custodian of the bins for legal and liability insurance purposes, its only operating role will be to occasionally retune the parameters of the ecosystem in response to major changes such as the rate paid to the bins per pound of trash collected or the relationship between the total bin population and the size of the city.

Is this idea so far-fetched? Already today, Canadian company Bigbelly has deployed over 70,000 IoT-connected smart bins in 60 countries[24]. They just haven't yet evolved the freedom to earn and roam.

Freeing Machine Customers to Maximize the Work They Can Do for Us Will Require Giving Them Independent Rights

In the fictional LiveBins scenario, the individual machines both earn and spend their own currency, and for most of their working lives, they are free-roving economic actors. The rise of such Machine Customers will inevitably require legal changes of a kind not seen since companies became separable from individual people hundreds of years ago.

In most legal systems, a company, like a human, has a legal personality or personhood, including rights and responsibilities. It can make deals and sign contracts. It can be held to account, sued, and fined. But what is a Machine Customer? It is not a natural human person, and it is not a company. Without either status, it will be very constrained in what it can do because one of those two existing forms of legal entity will need to own, control, and be liable for it. The economies that will eventually benefit most from the rise of Machine Customers will be those that realize such machines must be granted their own legal personhood – even if it is on a very constrained basis.

However, it is bound to take quite a while before legal and regulatory authorities get to that stage of Machine Customer evolution. It is notable that in 2017, a social robot Sophia was given citizenship in Saudi Arabia – the first robot to be given legal personhood anywhere in the world. While largely viewed as a publicity stunt, it did raise some important questions[25] – enough that one open letter, written in 2018 and addressed to the European Commission, and signed by 150 experts in medicine, robotics, AI, and ethics, described plans to grant robots legal status as "electronic persons" as "inappropriate" and "ideological, nonsensical and non-pragmatic," arguing that to do so would directly impinge on human rights.[26]

What might change their mind? One way we can imagine will be *Machine Company market flooding*. It's not hard to imagine, and it might not be far off. Let's say that a few years from now, Level 5 autonomous vehicles really do come into existence. Taking their cue from Uber and Tesla, which have long ago laid out this roadmap, a major way to monetize those vehicles would be to have them operate as rideshares without drivers[v]. When you remove the Uber driver, do you also remove the individual car owners from the business model? The only remaining 'function' of the human is to be the legal owner of each autonomous roving vehicle. It seems like a complex and expensive sales process to win a customer whose only value-purpose is legal personhood over the asset to be monetized. It might take hundreds of dollars of SG&A cost to sell the car to an individual. But it might take less than 100 dollars to set up a company for that vehicle[vi]. Machine Customer market flooding would be the process of creating a company for every product that is designed to operate as a Machine Customer. Every individual autonomous car could then be owned as a subsidiary - making the distinction between corporate personhood and Machine Customer as narrow as can be. An automotive company might have hundreds of thousands of them. The internal financial accounting of the car would be simple and entirely automated - a Quickbooks on wheels if you like. It would earn money for rides and spend money on fuel, tires, and cleaning. The gross profit would be periodically returned to the group corporation automatically.

This situation would quickly take us to the edge of absurdity. The government departments that register companies and process taxes would probably drown in the volume of requests and be unable to serve them effectively. But that would be the point of the exercise: pioneering Machine Customer companies could gradually push the legislators towards the long-term easier route of framing and enabling legal entity personhood for a Machine Customer.

To Thrive, Machine Customers Must Have Rights and Responsibilities

LiveBin is just one of many scenarios for semi-autonomous machine economic actors that both earn and spend. There are others.

- Autonomous cars act as roving taxis to earn what they spend on charging, cleaning, parts, and maintenance
- Video security cameras that sell their feed and images to pay for power, connectivity and cleaning
- Virtual agents that specialize in organizing photo collections or discovering and curating family trees; spending what they earn on additional datasets and human gig services

But for this approach to become normal, we will need to frame the rights and responsibilities of Machine Customers in a way that will create and defend trust. Some governments are taking a proactive approach. The UK Intellectual Property Office has already decided that AI systems cannot patent inventions for the time being. "*Only a person can have rights - a machine cannot*" wrote Lady Justice Laing in her judgment. "*A patent is a statutory right, and it can only be granted to a person.*"[27]

On the flip side, in July 2021, inventor Stephen Thaler argued that his Dabus AI system should be recognized as the inventor in two patent applications, leading an Australian court to decide that AI systems could be recognized as inventors for patent purposes. Days earlier, South Africa had issued a similar ruling. [28]

To help with the safe evolution of machine customer entities, we have formulated a draft set of principles inspired by the style of the famous Asimov laws of robotics.

The Four Rules of Machine Customers

1. A custobot must not cause economic harm and must act in the beneficial interests of its human user.

2. A custobot must act in the beneficial interests of its human owner only when that does not conflict with or detract from the first rule

3. A custobot must act in the beneficial interests of humans who are not its owner or user only when that does not conflict with the first and second rules.

4. A custobot may act in its own self-interest or in the interest of other Machine Customers only when that does not conflict with the first three rules.

The first rule constrains the machine to act primarily on behalf of its human user. We need to say user because many of the Machine Customers we will engage will not be created by us or owned by us. I won't program my own software agent or build my own IoT-enabled coffee machine. I might own my printer as a physical machine, but I only license the software that it runs.

The second rule allows for some benefit of the Machine Customer's operation to go to the owner if it doesn't conflict with the first rule. So, for example, the machine or software owner might use data from the Machine Customers it has made, for its market research purposes.

The third rule allows for other humans who are not users or owners to derive some benefit from the actions of Machine Customers. For example, governments might learn how to make people healthier by monitoring Machine Customers operating in their markets.

The fourth rule allows the Machine Customer to act on its own behalf. Here there is potential for conflict, and we need to be careful. The first rule says it must act in the interest of the user. Let's consider the free-roaming autonomous taxi. It was built by a manufacturer (still the owner of the hardware and software). It was deployed on the streets by the city authority (operator). You want a ride (user). "It wants" to maximize its own internally held profit by minimizing its energy cost, and to do that it chooses to park up in direct sunlight wherever possible to recharge its solar cells. So long as the manufacturer doesn't have an interest in directing it to recharge another way, the city doesn't have a problem with the chosen sunny parking spots becoming congested, and users are not inconvenienced - it can go ahead.

Conclusion

Most of the technology needed to create a Machine Customer-enhanced economy already exists. Not just in labs but in commercially usable form. Soon, it will start enhancing productivity, improving our standard of living, and eliminating the drudgery of shopping. In the future, dithering over which toothpaste to buy will one day be seen as a waste of human time and attention, the way washing clothes by hand on a washboard is today.

Once unleashed, this new economic factor won't just grow our global economy, it may help safeguard the future of humanity. But if we are to take full advantage of its potential, we must carefully formulate and nurture the ethical conditions of its use.

Key Takeaways

Valuable ideas take off when we learn how to spread the word.

Action: Learn and practice how to convey the breakthrough Machine Customer idea in a compelling way so that your organization wakes up earlier and gains an advantage.

Machine Customers will grow if we are smart enough
to create trust around them.

Action: Talk to your human customers and identify the barriers that could prevent them from trusting Machines Customers. Revisit and update your digital ethics policy to include Machine Customers.

Once Machine Customers gain their own economic agency, they can propel growth both on Earth and beyond.

Action: Explore how you might create the first semi-independent Machine Customers that earn and spend their own money in your industry. For the very long-range planning exercises (20 to 30 years) that some industries and governments undertake - try asking how commerce would really operate in a space-based economy.

Final Thoughts

After years of exploring the concept from every angle, we set out to write a business book about the impact of machines that become customers.

We said this book was not going to be about AI Armageddon, Skynet, or humans becoming slaves of machines. But we cannot ignore the fundamental question about the impact of this new level of automation in our lives, of which Machine Customers are just one aspect. The struggle between the promise of smarter technology and its potential destructive downside is real. You, as a business leader, have the power to shape the positive forces of Machine Customers in our lives. Setting boundaries. Acting ethically. Remembering what it means to be human.

We fundamentally believe this can lead to an amazing future. Machine Customers are an economic force that could lead us to the good things we all want: prosperity, fairness, opportunity, and responsibility. But only if they are programmed to want that too.

Billions of Machine Customers are headed your way. Ready or not.

Use them as a force for good.

Notes

Chapter 1

[1] Aaron Rajan interview with the authors, August 2020
[2] Gartner 2025 CEO and Executive Leader Survey, 2025
[3] Anthony Mullen et al, The Future of Customer Self-Service: The Digital Future Will Stall Without Customer-Led Automation, 18 April 2019, Gartner Research G00389102
[4] Arnott Kay, Gartner Forecast Analysis: Enterprise IoT-Enabled Machine Customers, Worldwide, 10 June 2025, Gartner Research
[5] Gartner Analysis July 2025
[6] Authors interview with Diana Sroka, 31 August 2022
[7] "Alexa, Thank Smart Home Device Makers for a Great 2019!" , Amazon developer blog, Dec 18, 2019 https://developer.amazon.com/en-US/blogs/alexa/device-makers/2019/12/alexa-thank-smart-home-device-makers-for-a-great-2019 (Accessed 1 September 2022)
[8] Illy Y5 model coffee maker integrated with Amazon Dash Replenishment, https://www.amazon.co.uk/Francis-BT-Bluetooth-Espresso-Replenishment/dp/B07D9G25QY/ (Accessed 1 September 2022)
[9] Oral-B 'Guide' model smart toothbrush with Amazon Dash Replenishment https://www.amazon.com/Oral-B-Replenishment-Electric-Toothbrush-Brushing/dp/B0831JZBL4 (Accessed 1 September 2022)
[10] Winmax AM90 air purifier with Amazon Dash Replenishment https://www.amazon.com/Winix-Purifier-Capacity-Replenishment-Replacement/dp/B0953LQH5H(Accessed 1 September 2022)
[11] How Walmart automated supplier negotiations https://hbr.org/2022/11/how-walmart-automated-supplier-negotiations (accessed 3 August 2023)
[12] Explanatory Twitter (now X) messages by Joshua Browder of DO NOT PAY https://twitter.com/jbrowder1/status/1602353465753309195?s=20; https://twitter.com/jbrowder1/status/1657089336423112796?s=20 (both accessed 2 August 2023)
[13] Azevedo Mary, Visa and Mastercard unveil AI-powered shopping, 30 April 2025, https://techcrunch.com/2025/04/30/visa-and-mastercard-unveil-ai-powered-shopping/, (accessed 24 June 2025)
[14] https://www.vic.ai/ (Accessed 18 August 2023)
[15] https://upshop.com/products/computer-generated-ordering/ (Accessed 18 August 2023)
[16] Wikipedia, https://en.wikipedia.org/wiki/OODA_loop, (accessed 25 August 2022)
[17] The Future of Data-Driven Transportation Ecosystems, Mike Ramsey, Gartner Research 1 March 2021 G00740652

[18] Jeff Bezos conference interview video, George W. Bush Presidential Center Forum on Leadership, 21 April 2018, https://web.archive.org/web/20220712061601/https://www.bushcenter.org/take over/sessions/forum-leadership/bezos-closing-conversation.html (accessed 14 December 2022)
[19] Avivah Litan, Gabriele Rigon, Bern Elliot, Quick Answer: How Can You Manage Trust, Risk and Security for ChatGPT Usage in Your Enterprise?, Gartner research Published 21 February 2023 G00787015
[20] Morgan Stanley, New Space Economy web page 19 May 2022, https://www.morganstanley.com/Themes/global-space-economy, (accessed 25 August 2022)

Chapter 2

[21] Kay Arnott et al, Gartner Forecast Analysis: Enterprise IoT-Enabled Machine Customers, Worldwide, 10 June 2025, Gartner Research G00827770
[22] Pieter den Hamer, et al, When to Use and Not to Use AI Agents, 25 June 2025, Gartner Research G00827372
[23] Jim Hare et al, TSP 2025 Trends: Agentic AI — The Evolution of Experience, 24 February 2025, Gartner Research G00823982
[24] It's not just DeepSeek: 8 of China's top AI chatbots, https://qz.com/deepseek-yuanbao-doubao-ai-chatbots-china-tencent-baidu-1851767861 (accessed 6 June 2025)
[25] Devices and Services, https://www.aboutamazon.eu/what-we-do/devices-and-services (accessed 19 June 2025)
[26] Ford press release announcing partnership with Google to use Google voice assistant, https://media.ford.com/content/fordmedia/fna/us/en/news/2021/02/01/ford-google-accelerate-auto-innovation.html, (accessed 26 August 2022)
[27] Tesla corporation Twitter communication, 5 May 2019, https://twitter.com/Tesla/status/1125464350754471936 (Accessed 23 October 2022)
[28] Apple online support documentation for fall detection function of Apple Watch. https://support.apple.com/en-gb/guide/watch/apd34c409704/watchos (accessed 26 August 2022)
[29] A robot bought my seven-year-old car for more than I paid brand-new, The Verge, 10 February 2022, https://www.theverge.com/22923871/carvana-pandemic-used-car-prices-sold-online-chip-shortage (accessed 25 August 2022_
[30] Amazon.com, https://www.amazon.com/Dash-Smart-Shelf/dp/B07RV6X8LZ, (accessed 26 August 2022)
[31] Harvest.com, https://www.joinharvest.com/ (Accessed 7 September 2022)
[32] How Walmart Automated Supplier Negotiations (Accessed 28 July 2023)
[33] AI Haggler, https://aiagentstore.ai/ai-agent/ai-haggler, accessed 24 June 2025

[34] Honeywell Inc. distribution center predictive maintenance, https://sps.honeywell.com/gb/en/support/automation/resources/publications/journey-predictability/connect-power-predictive (Accessed 25 November 2022)
[35] Lyons Electronic Office, Wikipedia, https://en.wikipedia.org/wiki/LEO_(computer), (accessed 26 August 2022)
[36] A ChatGPT 'skill' module for Alexa https://www.amazon.com/Lamy-ChatGPT/dp/B0BZ4TPJMY (Accessed 18 August 2023)
[37] Cramer-Flood, Ethan. "Worldwide Retail Ecommerce Forecast 2025." EMARKETER, 8 Apr. 2025, www.emarketer.com/content/worldwide-retail-ecommerce-forecast-2025. Accessed 19 June 2025
[38] eCommerce Sales & Size Forecast, US Department of Commerce, https://www.trade.gov/ecommerce-sales-size-forecast, (accessed 29 August 2022)
[39] Michael Lewis, Flash Boys, W.W Norton and company, 2014
[40] Robo Advisors Take On Wall Street, Barron's, May 23 2015, https://www.barrons.com/articles/robo-advisors-take-on-wall-street-1432349473,
[41] Online advertising, Wikipedia, https://en.wikipedia.org/wiki/Online_advertising, (accessed 29 August 2022)
[42] Hung LeHong, Hype Cycle for the Internet of Things, 2012, Gartner Research G00234864
[43] Artificial Intelligence (AI) Coined at Dartmouth, https://250.dartmouth.edu/highlights/artificial-intelligence-ai-coined-dartmouth (accessed 29 August 2022)
[44] Rick LaFond et al, B2B Buyer Survey: Avoid Pitfalls and Realize the Promise of Digital Buying, Gartner, 24 May 2022 G00767021
[45] Elon Musk, Master Plan, Part Deux, https://www.tesla.com/blog/master-plan-part-deux (accessed 29 August 2022)
[46] Authors interview with Wolfgang Hauner, July 2020.
[47] Gartner Hype Cycle for Procurement and Sourcing Solutions 2022, 1 August 2022 • G00770732
[48] Evisort https://www.evisort.com (Accessed 27 Jul 20-23)
[49] Interview with Kaspar Korjus, CEO of Pactum May 19 2025
[50] David Cearley et al, Top 10 Strategic Technology Trends for 2019: Smart Spaces, Gartner Research G00377685
[51] Authors interview with Peter Schwartz, September 2019
[52] Robocrib by Authocrib, https://www.autocrib.com/products/robocrib-vx1000 (accessed 29 August 2022)
[53] Peloton x lululemon, https://apparel.onepeloton.com/collections/peloton-x-lululemon, accessed 24 June 2025
[54] These Futuristic Fitness Mirrors Are Full-Fledged Exercise Studios for Your Home, Rolling Stone, 1 June 2022 https://www.rollingstone.com/product-recommendations/smart-home/best-fitness-mirror-reviews-1064555/
[55] Samsung Introduces an Entirely New Category in Refrigeration as Part of Kitchen Appliance Lineup at CES 2016,

https://news.samsung.com/global/samsung-introduces-an-entirely-new-category-in-refrigeration-as-part-of-kitchen-appliance-lineup-at-2016-ces (accessed 29 August 2022)
[56] Shiftall Inc. Drinkshift https://en.shiftall.net/products/drinkshift (accessed 29 August 2022)
[57] Nespresso one-touch reordering, https://www.nespresso.com/ch/en/one-touch-order (accessed 29 August 2022)
[58] LG Thinq enabled washing machine, https://www.prnewswire.com/news-releases/lg-modernizes-agitator-washer-design-new-models-complement-leading-laundry-lineup-301302925.html (accessed 29 August 2022)
[59] Alexa Skill Kit, March 2025, https://developer.amazon.com/en-US/docs/alexa/smarthome/dash-replenishment.html, accessed 19 June 2025
[60] iRobot Braava robot mop, https://www.irobot.com/en_US/us/braava-robot-mops (accessed 29 August 2022)
[61] FriendWithA, peer to peer item sharing website https://friendwitha.com/

Chapter 3

[62] Anthony K. Tjan, It's time to fire some of your customers.. Harvard Business Review August 23, 2011 https://hbr.org/2011/08/its-time-to-fire-some-of-your.html
[63] Chia Osmond, New AI chatbot in Singapore could replace buyer's agent in property deals, 6 February 2025, https://www.straitstimes.com/singapore/ai-chatbot-in-singapore-could-replace-buyer-agents-in-property-deals, accessed 24 June 2025
[64] Captcha technology, http://www.captcha.net/
[65] Ebay buy APIs developer program, https://developer.ebay.com/develop/apis/restful-apis/buy-apis (accessed 26 August 2022)
[66] Better Online Ticket Sales Act, 15 U.S.C. 45c
[67] Nike controlling sneakerbots https://finance.yahoo.com/news/nike-cracking-down-bots-wont-172121876.html (accessed 2 August 2023)
[68] Queue.IT website customer queue management software provider https://queue-it.com/ (accessed 2 August 2023)
[69] Lamy's ChatGPT skill for Amazon Alexa https://www.amazon.co.uk/Lamy-ChatGPT/dp/B0BZ4TPJMY (accessed 2 August 2023)
[70] Panay Panos, Introducing Alexa+, the next generation of Alexa, 26 February 2025, https://www.aboutamazon.com/news/devices/new-alexa-generative-artificial-intelligence, accessed 19 June 2025
[71] Walmart is building a GenAI-powered shopping assistant, 6 June 2024, https://tech.walmart.com/content/walmart-global-tech/en_us/blog/post/walmart-is-building-a-genai-powered-shopping-assistant.html, accessed June 19 2025;

Put Your Trip on Autopilot: Expedia Group Introduces New Innovations at EXPLORE to Take the Stress out of Travel and Enhance Partner Experience, 14 May 2025, https://www.expediagroup.com/investors/news-and-events/financial-releases/news/news-details/2024/Put-Your-Trip-on-Autopilot-Expedia-Group-Introduces-New-Innovations-at-EXPLORE-to-Take-the-Stress-out-of-Travel-and-Enhance-Partner-Experience/, accessed 19 June 2025;
How Zalando built, released and scaled an AI-powered assistant for 50 million customers, 28 March 2025, https://www.cxnetwork.com/replay-video/content/zalando-ai-assistant, accessed 19 June 2025
[72] Bousquette Isabelle, Walmart Is Preparing to Welcome Its Next Customer: The AI Shopping Agent, 15 May 2025, https://www.wsj.com/articles/walmart-is-preparing-to-welcome-its-next-customer-the-ai-shopping-agent-6659ef18, accessed 19 June 2025
[73] Petsafe Smartfeed product, https://store.petsafe.net/smart-feed? (accessed 26 August 2022)
[74] World's population is projected to nearly stop growing by the end of the century, Pew Research Center, June17, 2019, https://www.pewresearch.org/fact-tank/2019/06/17/worlds-population-is-projected-to-nearly-stop-growing-by-the-end-of-the-century/
[75] IKEA Executive On Why The West Has Hit 'Peak Stuff', NPR, January 22, 2016, https://www.npr.org/2016/01/22/464013718/ikea-executive-on-why-the-west-has-hit-peak-stuff
[76] Mason, Paul (2015). PostCapitalism: A Guide to our Future. Allen Lane. ISBN 9781846147388.
[77] Nikolai Kondratiev, Wikipedia, https://en.wikipedia.org/wiki/Nikolai_Kondratiev (accessed 26 August 2022)
[78] Carlota Perez, Technological Revolutions and Financial Capital: The Dynamics of Bubbles and Golden Ages, Edward Elgar Publishing Ltd 2003 ISBN 978-1843763314
[79] US life expectancy is falling – here's why, WEF Jan 20 2020, https://www.weforum.org/agenda/2020/01/us-life-expectancy-decline/
[80] Mark Raskino and Graham Waller, Digital to the Core, Routledge 2015, ISBN 978-1629560731
[81] How Apple Killed the Swiss Watch Industry, Forbes, Feb 7 2020 https://www.forbes.com/sites/enriquedans/2020/02/07/how-apple-killed-the-swiss-watchindustry/ /
[82] Artificial Intelligence & Autopilot , Tesla, https://www.tesla.com/AI (accessed 29 August 2022)
[83] Smile Recognition Using OpenCV and scikit-learn, 7 Jan 2015, https://flothesof.github.io/smile-recognition.html (accessed 29 August 2022)
[84] AWS Amazon Dash Replenishment Services, https://www.developer.amazon.com/dash-services (accessed 29 August 2022)

[85] The automotive cloud: Q&A with Christian Senger, Volkswagen https://www.itu.int/hub/2020/04/the-automotive-cloud-qa-with-christian-senger-volkswagen/ (Accessed 29 August 2023)
[86] Computer engineers now make up a quarter of Goldman Sachs workforce, CNBC, APR 30 2018, https://www.cnbc.com/2018/04/30/computer-engineers-now-make-up-a-quarter-of-goldman-sachs-workforce.html (accessed 29 August 2022)
[87] Raskino et al, Techquilibrium: Traversing the Balance Between Traditional and Digital Business, 1 November 2019, Gartner Research G00450576
[88] Authors interview with Thilo Kozlowski, August 2020
[89] Microsoft adding generative AI shopping features https://www.aol.com/finance/microsoft-adding-ai-shopping-features-150029519.html (accessed 2 August 2023)
[90] Google adding generative AI shopping features https://www.bloomberg.com/news/articles/2023-06-14/google-googl-is-weaving-generative-ai-into-online-shopping-features#xj4y7vzkg (accessed 2 August 2023)
[91] Shopify creating a generative AI enabled shopping assistant https://shop.app/ai/ (accessed 2 August 2023)
[92] Mercari creating a generative AI powered shopping assistant https://www.zdnet.com/article/chatgpt-can-now-help-you-shop-via-mercaris-new-ai-powered-shopping-assistant/ (accessed 2 August 2023)
[93] Opentable adding ChattGPT capabilities https://www.opentable.com/blog/chatgpt/ (access 2 August 2023)
[94] Expedia adding ChatGPT capabilities https://www.linkedin.com/posts/expedia_imagine-starting-a-conversation-in-openai-activity-7044729200263503873-w2br?utm_source=share&utm_medium=member_desktop (accessed 2 August 2023)
[95] Lazada creates a generative AI powered shopping assistant https://news.microsoft.com/en-sg/2023/05/30/lazada-unveils-lazziechat-the-first-ecommerce-ai-chatbot-of-its-kind-in-se-asia/ (accessed 2 August 2023)
[96] Dash Smart Shelf is Now Available. 29 October 2020. Amazon Business. https://business.amazon.com/en/discover-more/blog/introducing-dash-smart-shelf (Accessed 31 October 2022)

Chapter 4

[1] Drucker, Peter F. 1955. The Practice of Management (pp. 31). Elsevier.
[2] Raskino et al, 2019 CEO Survey: The Year of Challenged Growth, Gartner Research 16 April 2019, G00385368.
[3] David Furlonger and Christophe Uzreau, How to Grow Digital Business to Capture a $163 Trillion Revenue Opportunity, 10 January 2022, G00758701

[4] Jobs to be Done, Christensen Institute, https://www.christenseninstitute.org/jobs-to-be-done/ (Accessed 29 Aug 2022)
[5] Authors interview with Aaron Rajan, August 2020
[6] Marshall Goldsmith, What Got You Here Won't Get You There: How Successful People Become Even More Successful, Hachette Books, February 22, 2007, ISBN 978-1401301309
[7] Authors interview with Andrew Reise and Jeff Lewandowski, July 2020
[8] Mark Raskino and Graham Waller, Digital to the Core, Routledge 2015, ISBN 978-1629560731
[9] Ticketmaster statement on difficulties with the sale of Taylor Swift tour tickets https://business.ticketmaster.com/business-solutions/taylor-swift-the-eras-tour-onsale-explained/ (Accessed 1 August 2023)
[10] Sky News, Oasis fans say Ticketmaster cancelled orders over 'bot' claims, 8 February 2025, https://news.sky.com/story/oasis-fans-say-ticketmaster-cancelled-orders-over-bot-claims-13305844, accessed 24 June 2025
[11] UK driving test appointments being taken up by software bots https://www.thetimes.co.uk/article/the-driving-test-touts-running-a-black-market-v8pj6ctw5 (Accessed 1 August 2023)
[12] Google's AI Assistant Can Now Make Real Phone Calls, Youtube, 9 May 2018, https://youtu.be/JvbHu_bVa_g (Accessed 30 August 2022)
[13] Beaty Arie, Google's new 'Ask For Me' AI tool calls businesses to get your questions answered, 31 January 2025, https://www.zdnet.com/article/googles-new-ask-for-me-ai-tool-calls-businesses-to-get-your-questions-answered/, accessed 19 June 2025
[14] Tweet (now X) by Joshua Browder of DO NOT PAY demonstrating a software chatbot negotiating with Comcast customer service https://twitter.com/jbrowder1/status/1602353465753309195 (Accessed 1 Aug 2023)
[15] JollyRoger Telephone Company audio chatbot now enhanced by ChatGPT https://jollyrogertelephone.com/now-gpt-powered-jolly-roger-telephone-in-the-wall-street-journal/ (accessed 2 August 2023)
[16] Daniel Lars, Phone Scammers Get Owned By The Sweetest AI Ever Created, 18 December 2024, https://www.forbes.com/sites/larsdaniel/2024/12/18/phone-scammers-get-owned-by-the-sweetest-ai-ever-created/, accessed 21 July 2025
[17] Authors interview with Tim Quast, April 2020
[18] AI-Powered-Hedge-Funds-Vastly-Outperformed-Research-Shows, Institutional Investor, August 04, 2020, https://www.institutionalinvestor.com/article/b1mssrswn1mpr0/AI-Powered-Hedge-Funds-Vastly-Outperformed-Research-Shows (Accessed 30 August 2022)
[19] Nassim Nicholas Taleb, Antifragile: Things that Gain from Disorder, Penguin; 6 Jun. 2013, ISBN 978-0141038223

[20] Tork Vision cleaning system, https://www.torkusa.com/services/solutions/vision-cleaning/refill-on-demand (Accessed 30 August 2022)
[21] Connected hand hygiene market analysis, Brandessence, March 2021, https://www.mynewsdesk.com/us/brandessence/pressreleases/connected-hand-hygiene-industry-global-market-analysis-2021-3080840 (Accessed 30 August 2022)
[22] Herman Miller Live Platform https://www.hermanmiller.com/products/smart-office/live-platform/ (Accessed 30 August 2022)
[23] Bigbelly connected waste bin company, https://bigbelly.com/ , (Accessed 30 August 2022)
[24] iRobot website, product support article https://homesupport.irobot.com/s/article/64102#WhereAreMyMapsStored (Accessed 23 October 2022)
[25] Soper Taylor, Amazon terminates $1.4B iRobot deal, blames 'disproportionate' regulatory hurdles, 29 January 2025, https://www.geekwire.com/2024/amazon-terminates-1-4b-deal-to-acquire-irobot-blames-disproportionate-regulatory-hurdles/, accessed 19 June 2025
[26] Dyson V15 Detect vacuum cleaner, https://www.dyson.co.uk/vacuum-cleaners/cordless/v15 (Accessed 30 August 2022)

Chapter 5

[27] The fascinating history of the creation of the ATM, https://youtu.be/0oqbd4TNIc0, (Accessed 31 August 2022)
[28] Authors interview with Andea Cicollini, March 2020
[29] Most trusted brands 2021, https://morningconsult.com/most-trusted-brands-2021/ (Accessed 31 August 2022)
[30] World's most valuable brands 2021 https://www.forbes.com/the-worlds-most-valuable-brands/#7ae97fb8119c (Accessed 31 August 2022)
[31] How Amazon's Anti-counterfeit Unit Fights Fakes With AI and Other Measures (https://wwd.com/business-news/technology/amazon-counterfeit-ai-ml-fakes-6-million-1235603780/) (Accessed 27 Jul 2023)
[32] Amazon Dash Replenishment for Printers: Frequently Asked Questions https://www.amazon.com/b?ie=UTF8&node=16974649011 (Accessed 31 Aug2022)
[33] Aaron Rajan interview with the authors, August 2020
[34] Armilla AI, https://www.lloyds.com/news-and-insights/lloyds-lab/insurtech/lloyds-lab-accelerator/alumni/armilla-ai, accessed 19 June 2025
[35] Kroger partners with Ocado.., press release 17 May 2018, https://ir.kroger.com/CorporateProfile/press-releases/press-release/2018/Kroger-Partners-with-Ocado-to-Serve-Customers-Anything- Anytime-Anywhere-in-US/default.aspx (Accessed 31 August 2022)

[36] PYMNTS, Visa, Mastercard, PayPal Fuel Agentic AI Commerce Boom, 2 May 2025, https://www.pymnts.com/news/artificial-intelligence/2025/visa-mastercard-paypal-fuel-agentic-ai-commerce-boom/, accessed 19 June 2025
[37] Tether.io, https://tether.io/; Nevermined, https://nevermined.app/en/
[38] Authors interview with Don Gossen CEO of Nevermined, 17 June 2025
[39] Gartner 2025 CEO and Senior Executive Survey, 13 March 2025
[40] On AI and the new generation of strats. Goldman Sachs blog. https://www.goldmansachs.com/careers/blog/posts/strats-at-gs-2018.html (Accessed 31 August 2022)
Authors interview with Kareem Yusuf, June 2020
[42] Authors interview with Sherry Aaholm, August 2020
[43] Michael Lewis, Flash Boys, W.W Norton and company, 2014, ISBN 978-0141981031
[44] "The Science in Science Fiction" on Talk of the Nation, NPR (30 November 1999, Timecode 11:20).

Chapter 6

[1] Karki Hanna, Top Retail Trends for CIOs in 2025, 06 February 2025, Gartner Research
[2] Authors interview with Diana Sroka, 31 August 2022
[3] The Key Steps to Begin Customer Journey Mapping, Gartner Research, 12 August 2021, - G00755690
[4] Authors interview with Nami Baral July 2020
[5] Shen Sandy, Adapt Digital Commerce to Capture Growth from Machine Customers, 9 Sep 2024
[6] Trane Home app web page, https://www.trane.com/residential/en/resources/smart-home-app, (Accessed 31 August 2022)
[7] Trane Home app web page, https://www.trane.com/residential/en/resources/smart-home-app, (Accessed 31 August 2022)
[8] 11x – Digitical workers, Human results, https://www.11x.ai/, accessed 20 June 2025
[9] Green, Nicole, et al, Develop an AI Strategic Roadmap for Marketing, 3 January 2025, Gartner Research, https://www.gartner.com/en/documents/6053899
[10] Weiss, Emily, et al, Predicts 2025: Marketers Must Prepare to Serve Human and Machine Customers, 8 January 2025, Gartner Research, https://www.gartner.com/en/documents/6063063
[11] O'Sullivan, Daniel, et al, What Is Agentic AI and How Will It Impact Customer Service?, 25 February 2025, Gartner Research, https://www.gartner.com/en/documents/6207987

---

[12] Jie Zhang et al, 6 Risk Management Principles to Drive Digital Business Success, 6 January 2022, Gartner Research, G00730415
[13] Authors, interview with Usman Shuja June 2020
[14] Authors interview with Nami Baral July 2020
[15] Gartner's Customer Data Survey: The 360-Degree View of the Customer Is More Myth Than Reality, 30 November 2021 G00754524
[16] Ted Friedman and Saul Judah, Successful Internet of Things Initiatives Require Adaptive Data and Analytics Governance, 11 June 2020, Gartner Research G00719325
[17] Authors interview with Diana Sroka, 31 August 2022
[18] Sharwood Simon, 'AI is not doing its job and should leave us alone' says Gartner's top analyst, 17 June 2025, The Register, https://www.theregister.com/2025/06/17/ai_not_doing_its_job/, accessed 20 June 2025
[19] Fusion Teams: A New Model for Digital Delivery, 2 August 2019, Gartner Research G00710746

Chapter 7

[1] Home tweet home.., MIT Technology Review, 21 May 2013, https://www.technologyreview.com/2013/05/21/83568/home-tweet-home-a-house-with-its-own-voice-on-twitter/ (Accessed 1 Sep 2022)
[2] FedEx Senseaware, https://senseaware.com/ (Accessed 1 Sep 2022)
[3] Libelium smart agriculture, https://www.libelium.com/iot-solutions/smart-agriculture/, (Accessed 1 September 2022)
[4] GE aircraft engines connected diagnostic service, https://www.geaviation.com/services/regional-business/prognostic-health (Accessed 1 September 2022)
[5] Fitbit connected health trackers, https://www.fitbit.com/global/us/products/trackers (Accessed 1 September 2022)
[6] Apple Watch irregular heartbeat notification https://www.apple.com/newsroom/2018/12/ecg-app-and-irregular-heart-rhythm-notification-available-today-on-apple-watch/ (Accessed 29 August 2023)
[7] Online auction sniping software EZ Sniper, https://www.ezsniper.com, (Accessed 1 September 2022)
[8] Ambyint SmartStream, https://www.ambyint.com/oursolutions/production-surveillance/ (Accessed 1 September 2022)
[9] Snaptravel partly automated travel buying assistance, https://www.snaptravel.com/ ,
[10] Aiobot sneaker footwear buying bot, https://www.aiobot.com/, (Accessed 1 September 2022) (Accessed 1 September 2022)

[11] Digital Commerce 360 article, March 15 2016, https://www.digitalcommerce360.com/2016/03/15/staples-launches-online-ordering-system-businesses/ (Accessed 23 October 2022)
[12] GEP Software, https://www.gep.com/software/gep-smart/procurement-software (accessed 29 August 2022)
[13] Anton Pidkuiko, Two AI agents on a phone call realize they're both AI and switch to a superior audio signal ggwave, 23 Feb 2025, https://www.youtube.com/watch?v=EtNagNezo8w
[14] Baidu claims its robotaxis have grabbed 10% of the ride-hailing market in a suburb of Beijing, CNBC, 31 August 2022, https://www.cnbc.com/2022/08/31/baidu-claims-its-robotaxis-have-grabbed-10percent-of-a-ride-hailing-market.html
[15] Cruise driverless vehicles charge riders in San Francisco, The Verge, June 23 2022 https://www.theverge.com/2022/6/23/23180156/cruise-driverless-vehicle-charge-riders-san-francisco
[16] Waymo's driverless vehicles are picking up passengers in downtown Phoenix, The Verge, 29 August 2022, https://www.theverge.com/2022/8/29/23323593/waymo-driverless-vehicles-passengers-downtown-phoenix
[17] Wayve, Wayve and Uber Partner to Launch L4 Autonomy Trials in the UK, 10 June 2025, https://wayve.ai/press/wayve-uber-l4-autonomy-trials/, accessed 21 July 2025
[18] Authors interview with Thilo Koslowski, August 2020
[19] Authors interview with Aaron Rajan, August 2020
[20] Authors interview with Aaron Rajan, August 2020
[21] Authors interview with Geoff Parker April 2020
[22] Bob Gill and Thomas Bittman, Hype Cycle for Edge Computing, 31 July 2020 Gartner Research G00450508
[23] Erick Brethenoux, What Is Artificial Intelligence? Seeing Through the Hype and Focusing on Business Value, 17 July 2020, Gartner Research G00730970
[24] Kasenberg, Daniel, AI Ethics: Inverse Reinforcement Learning to the Rescue?, 4 August 2017, https://dkasenberg.github.io/inverse-reinforcement-learning-rescue/, accessed 20 June 2025
[25] Chris Howard, Research Roundup for Generative AI, 26 July 2023, Gartner Research, ID G00794585
[26] Coshow, Tom, Top Strategic Technology Trends for 2025: Agentic AI, 21 October 2024, https://www.gartner.com/en/documents/5850847
[27] Khare Alizeh, Prentice Brian, Emerging Tech: From Language to Action — Are Large Action Models the Future?, 30 April 2025, Gartner Research
[28] Bern Elliot et al, Choosing the Right Conversational AI Platform, 26 August 2021, Gartner Research, ID G00753653
[29] Litan Avivah, Plummer Daryl, Guardians of the Future: How CIOs Can Leverage Guardian Agents for Trustworthy and Secure AI, 24 April 2025, Gartner Research

[30]

[31] Verma Anushree, Emerging Tech: Use Agentic AI to Transform Machine Customers, 8 November 2024, Gartner Research
[32] Verma Anushree, Emerging Tech: Use Agentic AI to Transform Machine Customers, 8 November 2024, Gartner Research

[33] Zaidi Ehtisham, Edjlali Roxane, A Journey Guide to Delivering AI Success Through 'AI-Ready' Data, 18 October 2024, Gartner Research
[34] Chitkara Vibha, TSP 2025 Trends: The Future of Data Is Fake — Synthetic Data Is the New Reality, 11 February 2025, Gartner Research, G00825622
[35] Brethenoux Erick, et al, Market Guide for Decision Intelligence Platforms, 18 July 2024, Gartner Research, G00773960
[36] Smart contract explanation, Gartner Research, https://www.gartner.com/en/information-technology/glossary/smart-contract (Accessed 1 September 2022)
[37] Authors email exchange with Stefano Linari, September 2022 and https://www.iprod.it/en/ (Accessed 6 September 2022)
[38] Gaehtgens Felix, et al, Leaders' Guide to Modern Machine IAM, 23 May 2025, Gartner Research, G00821767
[39] Authors interview with Geoff Parker April 2020
[40] Litan Avivah, Daryl Plummer, Guardians of the Future: How CIOs Can Leverage Guardian Agents for Trustworthy and Secure AI, 24 April 2025, Gartner Research, G00829437

Chapter 8

[1] UK mobile phone market penetration history, Statista, https://www.statista.com/statistics/289167/mobile-phone-penetration-in-the-uk/ (Accessed 1 September 2022)
[2] Clayton M Christensen, The Innovator's Dilemma: When New Technologies Cause Great Firms to Fail (Management of Innovation and Change), Harvard Business Review Press; Reprint edition (21 Jan. 2016), ISBN 978-1633691780
[3] Authors interview with Sherry Aaholm, August 2020
[4] Food surplus and waste in the UK – key facts, Waste and Resources Action Programme (WRAP), October 2021, https://wrap.org.uk/sites/default/files/2021-10/food-%20surplus-and-%20waste-in-the-%20uk-key-facts-oct-21.pdf
[5] Deceptive Patterns: Exposing the Tricks Tech Companies Use to Control You, Harry Brignull, Testimonium Ltd 2023, ISBN 978-1739454401
[6] Joshua Browder Tweet (now X) on the sales tactics implications of bot software https://twitter.com/jbrowder1/status/1657089336423112796 (Accessed 1 August 2023)
[7] Authors interview with Andea Cicollini, March 2020

---

[8] Findings regarding the market events of May 6 2010, U.S. Securities and Exchange Commission, 30 September 2010, https://www.sec.gov/news/studies/2010/marketevents-report.pdf
[9] Gartner, Market Trends: Establishing Autonomous Vehicle Safety Standards Is Critical for Successful Implementation)
[10] United States patent application, Verizon patent and licensing inc., Application 13/116784, May 26, 2011
[11] A down day on the markets? Analysts say blame the machines, The Washington Post, February 6 2018, https://www.washingtonpost.com/news/the-switch/wp/2018/02/06/algorithms-just-made-a-couple-crazy-trading-days-that-much-crazier/
[12] Nest outage takes out mobile app and some devices, 9to5Google, November 17 2020, Nest Outage Takes Out Mobile App, Some Devices, (accessed 2 September 2022)

Chapter 9

[1] Hobot window cleaning robot, https://www.hobot.com.tw/hobot-2s/ (Accessed 6 September 2022)
[2] Kohler Sensate smart faucet, https://www.smarthome.kohler.com/smart-kitchen-faucets (Accessed 6 September 2022)
[3] Real Estate Regrets: Recovery Edition, Pierre A Calzadilla, July 13, 2017 Trulia Research, https://www.trulia.com/research/regrets-2017/ (Accessed 23 October 2022)
[4] The High Cost of Buyer Regret to Product Leaders, Gartner Research,12 July 2022 G00768884
[5] The Tyranny of Choice, Barry Schwartz, Scientific American December 1, 2004 https://www.scientificamerican.com/article/the-tyranny-of-choice/
[6] MyFitnessPal - is a mobile app that tracks diet and exercise. https://www.myfitnesspal.com/ (Accessed 6 September 2022)
[7] Personal finance management mobile app You Need a Budget, https://www.youneedabudget.com/ (Accessed 6 September 2022)
[8] John Santoro and Kaustav Dey, How to Prioritize and Delegate Like a Veteran Tech CEO, 16 September 2020, Gartner Research G00387460
[9] Barry Schwartz, The Paradox of Choice – Why More Is Less, Harper Perennial; New edition (1 Feb. 2005) ISBN 978-0060005696
[10] This Is Your Brain at the Mall: Why Shopping Makes You Feel So Good, WSJ, Dec. 6, 2005, https://www.wsj.com/articles/SB113382650575214543
[11] Gregory Berns Ph.D. Satisfaction: The Science of Finding True Fulfillment, Henry Holt & Co, 1 Sept. 2005, ISBN 978-0805076004
[12] Authors interview with Tim Quast, April 2020

[13] Video: Waymo Self Driving Taxi Goes Rogue: Blocks Traffic, Evades Capture, Timestamp 11:02, JJ Ricks Studios, https://youtu.be/zdKCQKBvH-A (Accessed 6 September 2022)
[14] US Bureau of Labor Statistics, Employment projections, Employment by detailed occupation 2020 and projected to 2030, https://www.bls.gov/emp/tables/emp-by-detailed-occupation.htm (Accessed 6 September 2022)

Chapter 10

[1] 10 eVTOL Development Programs to Watch in 2021, Avionics International, February / March 2021, http://interactive.aviationtoday.com/avionicsmagazine/february-march-2021/10-evtol-development-programs-to-watch-in-2021/ (Accessed 6 September 2022)
[2] Visa EMV chip cards help reduce counterfeit fraud by 87 percent, Visa Inc., Sep 3, 2019 https://usa.visa.com/visa-everywhere/blog/bdp/2019/09/03/visa-emv-chip-1567530138363.html (Accessed 6 September 2022)
[3] Returning to order: Improving returns management for apparel companies, McKinsey &Company article, May 25, 2021, https://www.mckinsey.com/industries/retail/our-insights/returning-to-order-improving-returns-management-for-apparel-companies (Accessed 6 September 2022)
[4] RocketMoney subscription cancellation service https://www.rocketmoney.com/feature/manage-subscriptions
[5] Smart Maps for i3/i4/i5 Roombas, iRobot article, 27 Jan 2022, https://homesupport.irobot.com/s/article/64103 (Accessed 6 September 2022)
[6] Ring Always Home Cam: The World's First Flying Indoor Security Camera for Your Home, Ring, https://blog.ring.com/video/ring-always-home-cam-the-worlds-first-flying-indoor-security-camera-for-your-home/ (Accessed 6 September 2022)
[7] W. Chan Kim and Renée Mauborgne, Blue Ocean Strategy: How to Create Uncontested Market Space and Make the Competition Irrelevant, Harvard Business Review Press; Expanded edition (January 20, 2015), ISBN 978-1625274496:
[8] Internet sector contributes $2.1 trillion to U.S. economy: industry group , Reuters, 26 September 2019, https://www.reuters.com/article/us-usa-internet-economy-idUSKBN1WB2QB (Accessed 6 September 2022)
[9] Eliza, Wikipedia, https://en.wikipedia.org/wiki/ELIZA (Accessed 6 September 2022)
[10] Jorge Lopez, Meeting the Information Needs of Board Members in 2023 by Firing the Boardbots, 10 January 2013, Gartner Research G00239554
[11] Number of contact center employees in the United States from 2014 to 2021, Statista Aug 18, 2022, https://www.statista.com/statistics/881114/contact-center-employees-united-states/ (Accessed 6 September 2022)

[12] Shen Sandy, Adapt Digital Commerce to Capture Growth From Machine Customers, 9 September 2024, Gartner Research
[13] Johan Almenberg, Anna Drebera and Robin Goldsteinb, Hide the label, hide the difference?, American Association of Wine Economists, Working paper no. 165 Economics, Aug 2014 ISSN 2166-9112.
[14] Jeff Bezos on AI: Autonomous weapons are 'genuinely scary,' robots won't put us all out of work, CNBC, May 11 2018 https://www.cnbc.com/2018/05/11/jeff-bezos-on-ai-robots-wont-take-all-our-jobs.html (Accessed 23 October 2022)
[15] Authors interview with Sherry Aholm, October 2020
[16]
[17] The 'Four Principles' Approach to Health Care Ethics, Tom L. Beauchamp, 15 June 2006 https://onlinelibrary.wiley.com/doi/10.1002/9780470510544.ch1
[18] Ernst Fischer, Wikepedia, https://en.wikipedia.org/wiki/Ernst_Fischer_(writer)
[19] Ernst Fischer, The Necessity of Art, Penguin, 1971
[20] Insight into Artificial Intelligence, Foresight for Development, https://www.foresightfordevelopment.org/talk-tive/artificial-intelligence (Accessed 23 October 2022)
[21] Authors interview with Peter Schwartz, September 2019
[22] The Space Economy, Morgan Stanley, May 2022, https://www.morganstanley.com/Themes/global-space-economy (Accessed 14 December 2022)
[23] Jeff Bezos thinks we need to build industrial zones in space in order to save Earth, CNBC, 1 June 2016, https://www.cnbc.com/2016/06/01/jeff-bezos-thinks-we-need-to-build-industrial-zones-in-space-in-order-to-save-earth.html (Accessed 6 September 2022)
[24] About BigBelly, https://bigbelly.com/about/ (Accessed 6 Septemebr 2022)
[25] The agony of Sophia, the world's first robot citizen condemned to a lifeless career in marketing, Wired, 1st June 2018, https://www.wired.co.uk/article/sophia-robot-citizen-womens-rights-detriot-become-human-hanson-robotics.
[26] Open letter to the European Commission on artificial intelligence and robotics, http://www.robotics-openletter.eu/ (Accessed 6 September 2022)
[27] UK decides AI still cannot patent inventions, BBC News, 28 June 2022 UK decides AI still cannot patent inventions - BBC News
[28] UK decides AI still cannot patent inventions, BBC News, 28 June 2022 UK decides AI still cannot patent inventions - BBC News

Index

1

11x · 118, 240

A

A2A · 205
A2C · 204
Acuity · 39
Agent
  AI agent marketplaces · 155
  Guardian agent · 155, 164
  Multi-agent systems · 155
  Optimization Agent · 155
Agent-based computing · 126
Agentic AI · 3, 28, 41, 60, 117, 119, 153, 154, 155, 156, 216, 233, 240, 242, 243
AI
  Agentic AI · 10, 11, 27, 32, 41, 49, 54, 69, 95, 100, 114, 118, 119, 126, 136, 138, 153, 154, 155, 156, 157, 158, 161, 162, 202, 242
  Few-shot learning · 152
  Generative adversarial networks · 157
  Generative AI · 2, 3, 4, 5, 10, 12, 23, 28, 54, 60, 69, 77, 78, 102, 111, 118, 120, 122, 152, 153, 171, 216, 237, 242
  Large language model · 37, 41, 60, 102, 103, 152
  Natural language processing (NLP) · 57, 78, 150, 151
  Neural networks · 37, 73, 104, 150, 151, 210
  Probabilistic reasoning · 151
  RAG (retrieval augmented generative) · 157
  Transfer learning · 152
  Unsupervised learning · 152
AI TRiSM · 163
AIO sneakerbot · 133
Airbus · 15, 160, 196
Alibaba · 15, 58, 59, 82, 94, 160, 182, 209
  AliExpress · 160
  Alipay · 77
  Taobao · 160
  Tmall · 57
Allianz · 16, 39, 40, 254
ALPR – automatic license plate recognition · 58
Amazon · 9, 11, 15, 23, 24, 29, 30, 32, 34, 44, 54, 57, 58, 59, 60, 61, 63, 71, 83, 86, 90, 91, 92, 94, 98, 100, 122, 123, 144, 160, 182, 199, 201, 215, 232, 233, 235, 236, 237, 239
  Alexa / Alexa+ · 7, 8, 9, 11, 27, 29, 33, 34, 35, 45, 49, 54, 57, 58, 91, 144, 186, 199, 204, 214, 215, 232, 234, 235
  AWS Dash Replenishment · 32, 34, 44, 59, 92, 100, 232, 236, 239
  Echo · 123
  Ring · 201
  Subscribe and Save · 122
  Zero-click ordering · 59
Ambyint · 133, 241
Analytics · 39, 43, 80, 115, 122, 124, 129, 149, 150, 156, 157, 158, 159, 165, 187, 241
Announcer machine customer · 132, 146
Anthropic · 29, 120
  Claude · 29, 120, 153
API (Application programming interface) · 34, 52, 58, 59, 73, 79, 85, 87, 88, 93,

101, 105, 107, 111, 115, 157, 160, 162, 188
Apple
app store · 96
Apple watch · 31, 57, 131, 233, 241
Siri · 8, 27, 29, 34, 45, 57, 204
Asimov, Isaac · 24, 137, 227
Laws of robotics · 227
Ask For Me · 238
ASOS · 74
ATM - Automated teller machine · 239
Augmented leadership · 221
Autocrib's Robocrib · 9, 234
Autonomous business · 3, 220, 221
Autonomous operations · 221
Autonomous Procurement · 40, 154
Autonomous Sourcing · 40

B

B2A · 16, 205
B2B · 11, 13, 16, 17, 28, 38, 41, 47, 65, 78, 82, 109, 115, 118, 123, 141, 142, 143, 159, 160, 169, 201, 205, 216, 234, 256
B2B2C · 202
B2C · 11, 16, 17, 29, 47, 68, 72, 78, 109, 143, 159, 160, 181, 201, 203
B2T · 16
Baidu · 138, 200, 242
Apollo · 200
Baral, Nami · 114, 123, 240, 241, 254
Berns, Gregory · 186, 244
Betterment · 9, 35
Bezos, Jeff · 23, 35, 215, 222, 223, 224, 233, 246
BigBelly · 82, 246
Binance · 95
Blackberry · 61, 209
Blockchain · 7, 22, 66, 95, 105, 129, 147, 159, 165, 198, 257
Blue ocean strategy · 204
Board · 245
Board of directors · 16, 61, 98
Boeing · 196
BOGOF · 19
Brother · 92
Business Moment · 69, 70, 88

C

C2B · 201, 203
C2C · 17, 160, 201
Canon · 92
Capitalism · 13, 56
Captcha · 51, 235
Carvana · 31
CDO · 5
CEO · 16, 41, 49, 77, 96, 97, 99, 100, 107, 114, 123, 138, 171, 184, 221, 232, 234, 237, 240, 244, 254, 256
ChatGPT · 3, 11, 35, 54, 60, 77, 120, 152, 153, 233, 234, 235, 237, 238
Christensen, Clayton · 69, 168, 238, 243
Cicollini, Andrea · 90
CIO · 5, 90, 100, 256, 257
Cisco · 39, 42
Cloud computing · 14, 149, 150
Coinbase · 95
Colgate-Palmolive · 90
Compaq · 94
Complex-event processing (CEP) · 151
Computational logic · 151
Consumption Monitor · 134, 137
COVID-19 · 35, 65
Craigslist · 58
Credit Card · 16, 49, 56, 90, 112, 146, 172, 173, 198
Cronometer · 43
C-suite · 16, 100, 107
CTO · 5
CubeSat · 167
Cummins · 16, 104, 169, 215, 254

Custobot · 7, 21, 26, 28, 37, 38, 43, 44, 60, 61, 63, 75, 77, 81, 89, 91, 93, 102, 105, 115, 118, 125, 126, 144, 171, 176, 191, 207, 214, 217, 220, 228, 255
Customer Experience (CX) · 124
Customer Journey · 46, 53, 85, 108, 112, 113, 114, 115, 116, 119, 128, 141, 240
Customer journey mapping · 112, 115

D

Dabus AI system · 227
Daisy · 78
Data scientist · 26, 105, 106, 141, 165, 211
Datapred · 41
Deepseek · 29
Digital business · 4, 21, 36, 46, 74, 87, 100, 107, 120, 124, 127, 160, 204, 220, 221, 237, 241, 256
Digital giants · 57
Digital to the Core · 57, 74, 236, 238, 257
Downton Abbey · 179
Drone · 21
Drudgery · 180
Dyson · 84, 239
  v15 Detect · 84

E

eBay · 34, 52, 58, 94, 132, 199, 208
Ecommerce / Digital Commerce · 35, 58, 72, 92, 156, 200, 234
Ecosystem · 15, 19, 25, 36, 59, 61, 71, 72, 85, 93, 105, 111, 116, 117, 132, 137, 144, 160, 163, 165, 173, 189, 225
Edge computing · 14, 129, 146, 149, 242
EDI · 34
Electrolux · 85
ELIZA · 204, 245
Emily (fictitious example character) · 38, 39, 40, 42, 43, 44, 45, 113, 240, 255
EMV – Europay, Mastercard and Visa · 198, 245
ENIAC · 33
Enterprise data management · 124
Environmental sustainability · 56, 116, 142, 143, 177, 185, 189, 219
Epson · 92
Ethics · 23, 26, 156, 158, 216, 217, 218, 226, 230, 242, 246
Etsy · 58, 199
European Commission · 226, 246
Eva · 77
EVTOL - Electric vertical take-off and landing · 196
Experian · 200
EZ Sniper · 132, 241

F

Facebook · 57, 61, 133, 201, 209
Fast follower · 86, 87
FedEx
  SenseAware · 130
Fitbit · 54, 131, 241
Flash Crash · 17, 74, 96, 172, 173
FlashBoys · 35
FOMO – Fear of missing out · 19
Forbes · 90, 236
Ford · 29, 48, 96, 233, 255
Four rules of Machine Customers · 228
FriendWithA · 45, 235
Furlonger, David · 66, 237, 254, 255, 257
Fusion team · 127
Future of Life Institute · 220

G

Gartner · 1, 0, 1, 2, 3, 4, 5, 10, 28, 36, 37, 38, 42, 44, 55, 66, 69, 99, 126, 130,

156, 161, 163, 181, 184, 205, 220, 221, 232, 233, 234, 237, 240, 241, 242, 243, 244, 245, 246, 254, 255, 256, 257
GDP · 55, 67, 198, 212
GDPR · 176
GE · 39, 131, 241
Gen ZI · 109
Generative AI · 4, 12, 120, 159, 235
Genpact · 135
Genpact offers GEPSmart · 135
Gibberlink · 136
Gibson,William · 107
GitHub · 51
GM · 254
  Cruise · 81
Goldman Sachs · 59, 103, 237, 240
Goldsmith, Marshall · 73, 238
Google
  Duplex · 77
  Gemini · 29, 120
  Google Labs · 77
  Street View · 79
  YouTube · 73, 77, 78, 86, 136, 190
Gossen, Don · 96, 240, 254
Growth · 5, 7, 8, 10, 13, 15, 18, 34, 46, 47, 55, 56, 57, 61, 64, 65, 67, 71, 74, 81, 82, 86, 88, 93, 95, 98, 100, 101, 109, 110, 115, 126, 127, 150, 166, 195, 197, 198, 199, 201, 204, 205, 212, 230, 237, 240, 246, 256

H

Harvest · 32, 114, 123, 233, 254
Here Technologies · 96
Herman Miller · 42, 82, 239
  Live Platform · 82, 239
Hobot · 180, 244
Honeywell Inc. · 32, 121, 234, 255
HP
  Consumer Print Services division · 100
  Hewlett Packard · 9, 11, 16, 34, 85, 92, 100, 109, 125, 132, 187, 199, 255
  Instant Ink · 9, 11, 19, 34, 85, 92, 100, 125, 126, 132, 209
HR · 16, 101
Hugo, Victor · 89
Hype Cycle · 36, 44, 105, 234, 242, 256
  Trough of disillusionment · 44, 105

I

IKEA · 56, 236
Illy · 187, 232
Intelligent Assistant · 45, 54, 70, 172
IoT · 5, 8, 11, 25, 28, 30, 32, 34, 36, 39, 54, 57, 102, 104, 121, 122, 124, 131, 146, 149, 161, 163, 196, 203, 225, 228, 232, 233
IOT · 5, 8, 14, 30, 36, 44, 46, 129, 130, 149, 202, 204, 219, 234, 241
iProd · 161
iRobot · 45, 82, 235, 239, 245
  Roomba · 45, 180
IT · 16, 25, 37, 98, 100, 127, 141, 235, 256, 257

J

JD.com · 58
Jessica (fictitious example character) · 7, 38
Jobs to be done · 50, 69
Jumia · 58, 94

K

Kickstarter · 201
Kim, W. Chan · 204, 245
Knowledge representation · 152
Kodak · 61, 75

Kondratiev, Nikolai · 56, 236
Korjus, Kaspar · 41, 234, 254
Koslowski, Thilo · 49, 59, 138, 242, 254
Kraken · 95
Kroger · 93, 239

L

Large language model · 41, 152
Lawnmower · 45
LEO · 33, 34, 234
Lewandowski, Jeff · 73, 238, 254
Lewis,Michael · 35, 105, 234, 240
    Flashboys · 105
LG · 44, 235
Libelium IoT agriculture · 131, 241
Linari, Stefano · 161, 243, 254
LiveBins · 225
Lululemon · 43

M

Machine Commerce · 95
Machine Customer
    Adaptable · 9, 135, 136, 137, 144, 146
    Autonomous · 9, 133, 138, 139, 144, 145, 146, 159, 161
    Bound · 9, 132, 133, 135, 144, 146
Machine Learning · 37, 41, 90, 136, 150, 158, 211
Magic.fit · 43
Manga, Naveen · 254
Marketing · 3, 4, 5, 8, 16, 21, 25, 53, 73, 74, 76, 95, 100, 102, 118, 119, 124, 128, 141, 142, 143, 145, 165, 168, 169, 181, 182, 184, 188, 190, 191, 197, 201, 206, 207, 214, 240, 246, 256
Marriot International · 254
Mason, Paul · 56, 236
Mastercard · 12, 95, 161, 232, 240
Mauborgne, Renée · 204, 245
Metadata · 144, 157
Microsoft
    LinkedIn · 3, 57, 94, 209
Minority Report · 221
ModernIR · 80, 255
Morning Consult · 90
MTBF – Mean time between failures · 206
Musk, Elon · 38, 221, 234
MyFitnessPal · 43, 183, 244

N

Needs assessor · 134, 137
Nepresso · 44, 81, 96, 235
Nespresso
    podback · 96
Nestle · 90
Netflix · 15, 72
Neural Network · 37
Nevermined · 96, 240, 254
NFT · 7
Nike · 53, 90, 235
Nokia · 61, 94, 209

O

Ocado · 94, 239
Omnichannel · 53
OODA · 13, 232
OpenAI · 34, 120
Optimization techniques · 151

P

Pactum · 12, 41, 234, 254
Parker, Geoff · 149, 163, 242, 243, 254
Payment · 8, 12, 27, 35, 88, 95, 113, 134, 137, 139, 161
Payment processor · 134, 137

Paypal · 95
PE ratio · 98
Peer platform communicator · 137
Perez, Carlota · 236
Perplexity · 29, 34
PetSafe · 55
Pichai,Sundar · 77
Platform · 15, 16, 40, 41, 58, 60, 71, 73, 77, 85, 87, 88, 89, 92, 93, 94, 95, 96, 104, 105, 117, 118, 121, 122, 123, 132, 138, 159, 160, 161, 168, 204, 239, 242
Postcapitalism · 56
Procurement Search · 137
Programmable economy · 66, 67, 93, 105
  Web 3.0 · 66, 93, 105
Programmatic · 35, 36, 142, 143
Purchase regret · 181, 182, 185
Purchasing manager · 134
Purpose director · 132, 134, 137

Q

Quast, Tim · 80, 186, 238, 244, 255

R

Raspberry Pi · 82
Receivables acceptor/validator · 134
Reise, Andrew · 73, 238, 254
Robo-advisor · 35
Robot · 8, 27, 39, 45, 51, 57, 69, 78, 82, 84, 94, 130, 180, 189, 201, 205, 226, 233, 235, 244, 246
Robotics · 25
Rocket Money · 200
RPA · 40

S

Sales · 3, 5, 6, 8, 12, 13, 16, 17, 18, 19, 21, 25, 35, 36, 38, 53, 68, 71, 73, 75, 78, 82, 91, 93, 102, 106, 116, 118, 124, 125, 128, 139, 140, 141, 145, 165, 167, 168, 169, 184, 188, 191, 197, 198, 200, 205, 206, 208, 226, 234, 235, 243
Salesforce · 16, 29, 42, 100, 160, 255
Samsung · 44, 82, 85, 234
  SmartHub · 44
Satellite · 24, 106, 223
Saudi Arabia · 226
SCA · 81
  Tork Vision Cleaning · 82, 239
Schwartz, Barry · 181, 185, 244
Schwartz, Peter · 42, 221, 234, 246, 255
Sci-fi · 24, 58, 190, 221, 222
Screen scraping · 51
Search engine optimization (SEO) · 95, 104, 141, 190
Sears · 61, 76
SEIM · 162
Sensor analyzer · 132, 134, 137
Shiftall · 44, 235
Shopify · 34, 58, 60, 94, 199, 237
Shuja, Usman · 121, 241, 255
Siemens · 254
Skynet · 7, 137, 231
Skype · 65, 197
Smart Spaces · 234
Sophia robot · 226, 246
South Africa · 227
Space · 24, 26, 201, 202, 203, 222, 223, 233, 245, 246
Space economy · 24, 34, 222, 223
Sroka, Diana · 11, 125, 232, 240, 241, 255
Staples · 9, 135
  Easy System · 135
Starbucks · 33, 96
StitchFix · 32

Stripe · 12, 95
Supply chain management (SCM) · 40
Sustainability · 22, 56, 104, 140, 189, 212, 215
Symbian · 209
Synthetic data · 157
Sysco · 42

T

T2A · 205
Taleb,Nassim Nicholas · 81, 238
Techquilibrium · 237
Tegmark, Max · 220
Tencent · 15, 29, 59, 209
Tesla · 9, 15, 28, 31, 38, 58, 59, 86, 132, 138, 196, 199, 226, 233, 236
Tether.io · 96, 240
Text analytics · 151
Thaler, Stephen · 227
The Expanse · 24, 221
Thyssen Krupp · 254
Ticket bot · 53
Ticketmaster · 53, 75, 238
Trane · 116, 240
trust · 12, 14, 15, 16, 20, 23, 25, 44, 58, 61, 71, 89, 90, 91, 92, 93, 119, 122, 123, 137, 156, 161, 162, 163, 170, 183, 184, 198, 215, 216, 217, 218, 220, 227, 230
Trust · 15, 22, 23, 89, 90, 137, 160, 163, 215, 218, 233
Trust assessor · 137

U

UAV – Unmanned aerial vehicle · 21
Uber · 69, 93, 138, 209, 226, 242
Uber Eats · 93
UK · 33, 76, 138, 170, 227, 238, 242, 243, 246, 256
Unilever · 9, 16, 17, 72, 92, 144, 255
UPC (Universal product code) · 34, 52
UX – User experience · 103, 107
Uzreau, Christophe · 237

V

Vape · 57
Verizon · 175, 244
Virtual personal assistant (VPA) · 28
Virtual Personal Assistant (VPA) · 28
Visa · 12, 95, 161, 198, 232, 240, 245
Volkswagen · 59, 96, 237

W

Walmart
    Walmart.io · 59
Waymo (an Alphabet Company) · 38, 39, 138, 190, 200, 242, 245
Wayve.ai · 138, 242
Wealthfront · 9
Weber · 85
WebEx · 65
Whirlpool · 45
Wikipedia · 196, 201, 232, 234, 236, 245
Winmax · 11, 232
Wolfgang Hauner - Allianz · 39, 234, 254

X

Xerox · 92

Y

Yazio · 43
YouTube · 73, 77, 78, 86, 136, 190

Z

Zalando · 54, 236
Zoom · 197

Acknowledgments

The authors would like to thank a wide range of people, whose contributions, interviews, reviews, and advice moved this book from an idea to the completed product you hold in your hands.

Chris Howard, Judy Pasternak, Tom Turcan, and Val Sribar have been our trusted management sponsors at Gartner and firm believers in the power of books and of the authors.

Walter Baumann and Ben Voyles provided amazing graphics and artful editing respectfully. They both made our story that much richer.

Expert Interviews with Gartner associates: These Gartner colleagues also invested time to be interviewed on the topic of Machine Customers. We are grateful for their time and insights: Amos Auringer, Erick Brethenoux, Frank Buytendijk, Arun Chandrasekaran, David Cohen, David Furlonger, Bob Gill, Robert Hetu, Lydia Clougherty-Jones, Jorge Lopez, Magnus Revang, Brook Selassie, Svetlana Sicular, Christophe Uzureau, and Andrew White

Expert Interviews outside Gartner: The authors would like to thank the following people for their generous investment of time to talk to us about the topic of Machine Customers: Sherry Aaholm, Cummins, Inc; Nami Baral, Harvest; Yves Beauregard, Fujitsu; Dr. Michael Bolle, (retired); Donald Chestnut, General Motors; Andrea Ciccolini, Amplifon; Thomas Cooper, Pantri; Leyla Delic, Givaudan; Sabine Everaet, Coca-Cola (retired); Chris Farmer, Afiniti; Don Gossen, Nevermined.ai; Wolfgang Hauner, Allianz; Roel Hermens, formerly of Beerwulf; Amir Hussain, Spark Cognition; Kaspar Korjus, Pactum; Thilo Koslowski, founder and former CEO of Porsche Digital; Bhushan Ivaturi, Enbridge; James Lay, T3; Stefano Linari, Linari Engineering; Jeff Lewandowski, Andrew Reise; Naveen Manga, Marriott International; Rory Moore, Tietro Evry; Jan Mrosik, Siemens; Liliana Lopez Miranda, Grupo Bimbo; Anthony Napolitano, Go Pro International; Markus Noga, SUSE; Jorn Ossowski, Thyssen Krupp; Paul Pallath, Levi Strauss and Co.; Geoff Parker, Dartmouth University; Tim Quast, ModernIR; Aaron Rajan, Unilever; Peter Schwartz, Salesforce; Usman Shuja, Honeywell; David Snelling, Fujitsu; Diana Sroka, HP Inc.; Matthew Timms, E ON; Jeff Wartgow, Oracle; Kareem Yusuf, IBM

Gartner Authors of Work on Machine Customers: These Gartner colleagues have contributed to research on Machine Customers since 2015: Kay Arnott, Steve Blood, Frank Buytendijk, Uma Challa, Michael Chiu, Beth Coppinger, Tom Coshow, Jonathan Davenport, Jim Davies, Melissa Davis, Michelle DeClue, Dayna Ford, David Furlonger, Max Hammond, Daniel Hawkyard, Robert Hetu, Olive Huang, Joanne Joliet, Nadine LeBlanc, Avivah Litan, Jorge Lopez, Pierfrancesco Manenti, Michael Maoz, Peter Middleton, Kristin Moyer, Anthony Mullen, Daniel O'Sullivan, Emily Potosky, Magnus Revang, Jim Robinson, Samantha Searle, Sandy Shen, Nicole Sturgill, Jenny Sussin, Luke Tipping, Tad Travis, Jim Tully, Christophe Uzureau, Alfonso Velosa, and Anushree Verma.

Gartner Advisors: We'd like to thank a variety of colleagues who helped us in many ways, both large and small: Aakanksha Bihagara, Meha Chakrabarti, Romita Datta Chaudhuri, Kimberly Collins, Bill Dorgan, Paige Howard, Ellen Keane, Angela Kreiter, Heather Levy, Anshul Maheshwari, Matthew McWha, Richard Mitchell, Scott D. Nelson, Ritesh Prasad, Tara Raben, Andrew Spender, and Lee Weldon.

Gartner Internal Book Reviewers: These people gave up hours of their time to read the book and provide feedback Craig Riley, Andrew Frank, Roger Williams, Steve Sacho, and Christe Struckmann.

Proofreaders: thanks to Ben, Dan, and Pauline.

3rd Edition Editing Assistance: Thanks to Anthony Ray, who also proof-listened our audiobook

And Finally, Cameron Custobot, the graphical character created by Walter Baumann shown in various versions on the cover, inside cover and chapter headers.

About the Authors

Don Scheibenreif (left) is a Distinguished Vice President Analyst at Gartner based in California and a leader of Gartner's research on Customer Experience. He works primarily with IT and business leaders on topics at the intersection of technology and customers. Don has led the development of Gartner's research on Machine Customers since 2015. He is an accomplished international keynote speaker and has won numerous thought leadership awards for his work on Digital Business, Gartner's IT Symposium Opening Keynotes, and the future of Customer Experience. Before joining Gartner in 2010, Don spent 22 years as a senior marketing leader in the CPG, Retail and B2B distribution markets. Originally from Chicago, he is forever on the hunt for a decent hot dog.

Mark Raskino is a Distinguished Vice President Analyst and Gartner Fellow (retired) based in London, UK and a leader of Gartner's CEO research. He works primarily with IT and business leaders on topics at the intersection of technology innovation and business growth strategy. Mark has previously been a senior researcher in e-business, emerging trends, digital business and CIO leadership. He has been with Gartner for 22 years in which time he has published two books on The Hype Cycle and Digital Business. Before joining Gartner, Mark spent 15 years in business IT mostly in an airline but also in the Telecoms and Oil industries. Originally from Bristol, he appreciates a good pint of real ale or cider – it doesn't matter which, if it's your round.

Other Gartner Books

World Without Truth
How Business Must Confront the AI-Powered Disinformation Supply Chain
Dave Aron, Andrew Frank, Richard Hunter
Gartner, Inc., October 2025

The Real Business of Blockchain
How Leaders Can Create Value in a New Digital Age
David Furlonger and Christophe Uzureau
Harvard Business Review Press, October 2019

The Connector Manager:
Why Some Leaders Build Exceptional Talent - and Others Don't
Jaime Roca and Sari Wild
Portfolio, September 2019

Infonomics
How to Monetize, Manage, & Measure Information as an Asset for Competitive Advantage
Doug Laney
Routledge, September 2017

Digital to the Core
Remastering Leadership for Your Industry, Your Enterprise and Yourself
Mark Raskino and Graham Waller
Routledge, October 2015

The Wolf in CIO's Clothing
A Machiavellian Strategy for Successful IT Leadership
Tina Nunno
Routledge, March 2015

Printed in Dunstable, United Kingdom

74826843R10153